In Situ Hybridization Histochemistry

Editor

Marie-Françoise Chesselet, M.D., Ph.D.

Department of Pharmacology
School of Medicine
University of Pennsylvania
Philadelphia, Pennsylvania

CRC Press
Boca Raton   Ann Arbor   Boston

Library of Congress Cataloging-in-Publication Data

In situ hybridization histochemistry/ editor, Marie-Françoise Chesselet
    p.   cm.
    Includes bibliographies and index.
    ISBN 0-8493-6912-6
    1. Nucleic acid hybridization. 2. Histochemistry. 3. Nucleic acid probes. 4. Gene
expression—Research—Methodology.
I. Chesselet, Marie-Françoise.
QP620.I5 1990   574.87'328—dc20   89-71300 CIP

Direct all inquires to CRC Press, Inc., 2000 Corporate Blvd., N.W., Boca Raton, Florida, 33431.

© 1990 by CRC Press, Inc.

International Standard Book Number 0-8493-6912-6

Library of Congress Card Number 89-71300
Printed in the United States

INTRODUCTION

The technique of *in situ* hybridization histochemistry has become increasingly popular with scientists in a number of fields over the last few years. The unique ability of this method to allow for the detection of specific messenger RNAs in single cells makes it a method of choice to study the regulation of gene expression in small or heterogenous tissue samples. Technical advances have greatly increased the scope of *in situ* hybridization histochemistry and made it accessible to a variety of investigators with minimal expertise in molecular biology. The goal of this book is to present and discuss some widely used methods to perform *in situ* hybridization histochemistry, illustrate their potential with chosen examples and provide an update on some of the newest developments in the field. Most examples are drawn from the field of neurobiology, but the principles developed have much wider applications. It is not our intention to cover every study or method related to the use of *in situ* hybridization histochemistry, but to provide enough information to allow investigators to apply this new approach to a particular scientific question.

In the first two chapters, Lewis and Baldino and Bloch discuss the critical issue of probe choice and preparation, including the most recent advances in the development of nonradiolabeled probes. Jordan then provides detailed information on the different requirements for *in situ* hybridization in cells and tissue sections, and illustrates both approaches in studies of myelin gene expression in the central nervous system. The potential of *in situ* hybridization histochemistry for the study of normal and abnormal gene expression in the brain is further illustrated in the chapters by Frantz and Tobin and by Murray on the use of the technique to study mutant mice and mRNAs encoding growth factors and oncogenes.

The chapters by Eberwine et al. and by Soghomonian describe some of the newest and most promising developments of *in situ* hybridization histochemistry; *in situ* transcription and electron microscopic detection of mRNAs in tissue sections. The first method is expected to allow not only for the localization of mRNAs, but also for the determination of the translational state of specific mRNAs. The second method will provide unique information on the location of various mRNAs within the cell.

As illustrated in these chapters, *in situ* hybridization histochemistry has been most useful in determining the pattern of specific gene expression in a number of tissues. It is now clear from work in tissue homogenates and cell cultures that a number of factors, including developmental events, neuronal activity, and pharmacological treatments can modify gene expression, resulting in a change in the level of specific mRNAs. In order to study these phenomena with the anatomical resolution characteristic of *in situ* hybridization histochemistry, it will be necessary to develop a means of quantifying the results obtained with this method. This crucial question is addressed by Smolen and Beaston-Wimmer in Chapter 8. Several laboratories have now reported changes in mRNA levels in identified cell populations under a variety of experimental conditions. One may expect that future use of quantitative *in situ* hybridization histochemistry will greatly contribute to our understanding of factors regulating gene expression in complex tissues. Together with the development of more refined ways to use *in situ* hybridization for the localization of mRNAs into cells, improvement of quantification methods will most certainly broaden the use of *in situ* hybridization histochemistry in the near future. It is our hope that the methodological information and data reported in this volume will help and encourage more investigators to use the potential of *in situ* hybridization histochemistry to answer critical biological questions.

THE EDITOR

Marie-Françoise Chesselet, M.D., Ph.D., is Associate Professor in the Department of Pharmacology, University of Pennsylvania, Philadelphia. She received her Diplome de Docteur en Medeine (M.D.) from the University of Paris in 1974 and the Doctorat d'Etat es Sciences Naturelles (Ph.D.) from the same institution in 1978.

Dr. Chesselet's first appointment was as Assistant de Recherches at the Centre National de la Recherche Scientifique, Paris, after graduation from the University, and became Chargee de Recherches in 1981. She accepted a position as a Visiting Scientist at Massachusetts Institute of Technology, Cambridge, in 1982, and left to join the National Institute of Mental Health, Bethesda, MD in 1984. Prior to her current position, she was an associate professor of pharmacology and of anatomy at the Medical College of Pennsylvania, Philadelphia.

She has been the recipient of several grants from the National Institute of Mental Health, the National Science Foundation, and several private foundations, and has received honors including a NATO Fellowship, a Fogarty Postdoctoral Fellowship, and a Huntington's Disease Foundation of America Fellowship. She served on the Scientific Advisory Board of the Hereditary Disease Foundation and the Science Council of the Huntington Disease Society of America, and has been Councillor and President of the Philadelphia Chapter of the Society for Neuroscience.

Dr. Chesselet has membership in the American Society for Neurosciences and the International Narcotic Research Council and has published elsewhere extensively.

to

Allan J. Tobin
Michael J. Brownstein
Hans-Urs Affolter

and all who helped us get started

CONTRIBUTORS

Frank Baldino, Jr., Ph.D.
Research Director
Cephalon, Inc.
West Chester, Pennsylvania

Patricia Beaston-Wimmer
Department of Anatomy
Medical College of Pennsylvania
Philadelphia, Pennsylvania

Bertrand Bloch, M.D., Ph.D.
Professor of Histology
Laboratoire d'Endocrinologie
Universite de Bordeaux II
Bordeaux, France

Marie-Françoise Chesselet, M.D., Ph.D.
Associate Professor
Department of Pharmacology
University of Pennsylvania
Philadelphia, Pennsylvania

James Eberwine, Ph.D.
Department of Pharmacology
University of Pennsylvania
Philadelphia, Pennsylvania

Gretchen Frantz
Department of Biology
University of California
Los Angeles, California

Craig A. Jordan, Ph.D.
Senior Staff Fellow
Laboratory of Viral and Molecular
  Pathogenesis
NIH/NINDS
Bethesda, Maryland

Michael E. Lewis, Ph.D.
Senior Scientist
Cephalon, Inc.
West Chester, Pennsylvania

Marion Murray, Ph.D.
Professor of Anatomy
Medical College of Pennsylvania
Philadelphia, Pennsylvania

Arnold J. Smolen, Ph.D.
Associate Professor
Department of Anatomy
Medical College of Pennsylvania
Philadelphia, Pennsylvania

Jean-Jacques Soghomonian, Ph.D.
Department of Pharmacology
University of Pennsylvania
Philadelphia, Pennsylvania

Allan J. Tobin, Ph.D.
Professor
Department of Biology
University of California
Los Angeles, California

TABLE OF CONTENTS

Chapter 1

PROBES FOR *IN SITU* HYBRIDIZATION HISTOCHEMISTRY

Michael E. Lewis and Frank Baldino, Jr.

TABLE OF CONTENTS

I. INTRODUCTION

Prior to the development of *in situ* hybridization histochemistry, pedigree analysis or somatic cell genetics was required to estimate the location of genes on chromosomes. However, these traditional methods were superceded by the discovery that biosynthesized, radioactively labeled RNA could be used for hybridization to homologous DNA in cytological preparations of chromosomes to indicate the locus of the corresponding gene.[1-3] This technology was rapidly exploited to determine the chromosomal localization of genes encoding 18 and 28S ribosomal RNA, transfer RNA, and histone messenger RNA (mRNA) in a wide variety of species,[4] but was limited by the unavailability of probes for many sequences of interest. This limitation was overcome by the introduction of recombinant DNA technology, which made available a wide variety of complementary DNA (cDNA) probes of known sequence, which were then used to map single copy genes on chromosomal preparations.[5]

The introduction of recombinant cDNA probes also greatly facilitated the cytological study of mRNA, which had previously required the isolation of genome templates for the synthesis of radiolabeled cDNA probes,[6,7] a procedure with very limited applicability. This chapter is devoted to a discussion of the various types of probes which are now available for the study of mRNA *in situ,* as well as the use of various radioactive and nonradioactive probe labeling methods.

II. TYPES OF PROBES

A. cDNA Probes

1. Isolation and Labeling of cDNA Probes

Until recently, hybridization probes were almost invariably obtained from cloned pieces of DNA which are complementary to a particular mRNA species. The complementary DNA (cDNA) clones must be isolated from cDNA libraries of clones which are prepared by enzymatic reverse transcription of isolated mRNA into cDNA copies which are then made double-stranded and inserted into appropriate cDNA cloning vectors.[8] After the positive clone is identified (e.g., by oligonucleotide probe hybridization or antibody binding if an expression system is employed), *Escherichia coli* containing the recombinant plasmid with a cDNA copy of the relevant mRNA are grown in large quantities. The plasmid is then extracted, purified, and digested with a restriction endonuclease to excise the cloned DNA from the vector sequence. The DNA fragment is then purified by gel electrophoresis, eluted, and then labeled by nick translation,[9] i.e., using DNase I to generate random nicks and DNA polymerase I to initiate DNA synthesis with radioactive nucleotide triphosphates at the nick sites (Figure 1). Alternatively, mixed sequence hexadeoxynucleotides can be used as "random primers" to prime the synthesis (by the Klenow fragment of DNA polymerase I) of labeled DNA probes from restriction fragments which have been purified by

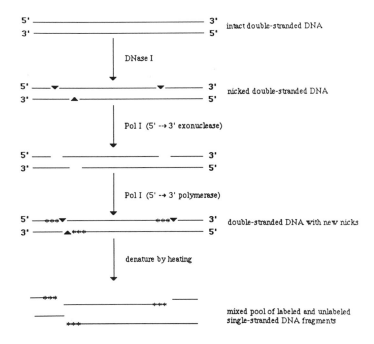

5' ──────────────────────── 3' intact double-stranded DNA
3' ──────────────────────── 5'

↓ DNase I

5' ──▼────────────▼───── 3' nicked double-stranded DNA
3' ────────▲──────────── 5'

↓ Pol I (5' → 3' exonuclease)

5' ── ────────── ──── 3'
3' ───── ────────────── 5'

↓ Pol I (5' → 3' polymerase)

5' ──✳✳✳▼──────────✳✳✳▼── 3' double-stranded DNA with new nicks
3' ────────▲✳✳✳──────── 5'

↓ denature by heating

──✳✳✳ ──────────────── ✳✳──
────── ──────✳✳✳────
✳✳✳─────────────────   mixed pool of labeled and unlabeled
single-stranded DNA fragments

FIGURE 1. Complementary DNA probe labeling by nick-translation (see text for further explanation). Note that method generates labeled and unlabeled fragments of various sizes which can reassociate.

agarose gel electrophoresis following restriction nuclease digestion.[10,11] The radioactive DNA fragments are then purified from unincorporated nucleotide by phenol-chloroform extraction or column chromatography.

2. Advantages and Disadvantages of cDNA Probes

DNA probes prepared as described above will contain many radioactive nucleotides and so can serve as usable probes for *in situ* hybridization, as detailed elsewhere.[12-14] However, as noted before,[15] there are some disadvantages associated with their use, including: (1) difficulty in obtaining the clones from recalcitrant investigators; (2) poor tissue penetration due to excessive probe length; (3) reannealing of the sense and antisense strands during hybridization, effectively reducing probe availability; (4) variable lengths, which precludes Tm studies; (5) the unavailability of particular DNA sequences due to lack of an appropriate restriction site; and (6) the need to establish microbiological and molecular biological methods, which may be particularly daunting to the histochemist who only wants to obtain probes to use as ligands to detect mRNA in tissue sections. Although denatured double-stranded cDNA probes should form hyperpolymers (partially reassociated fragments) which enhance the hybridization signal, such reaggregation appears instead to impair probe penetration to target mRNAs.[16] Some of these difficulties have been overcome by the introduction of a method for synthesizing single-stranded DNA probes from recombinant templates inserted in phage M13 vectors.[17] In this method, an M13 universal primer is used to initiate the synthesis of a radioactively labeled DNA strand which is then purified from the larger template molecule by restriction digestion and gel electrophoresis. While this procedure produces high specific activity single-stranded DNA probes which are usable for *in situ* hybridization,[18] the utility of the method is limited by the low efficiency of transcription (one transcript per template molecule) and possible contamination with vector sequence transcripts.[16] Because of these limitations, investigators have sought more efficient methods of probe synthesis.

B. cRNA Probes
1. cRNA Synthesis

For the synthesis of RNA, plasmids containing specific RNA polymerase promoter sequences (e.g., from phage T7 or the Salmonella phage Sp6) have been prepared with a multiple cloning site (i.e., polylinker) adjacent to the promoter, into which cDNA restriction fragments can be inserted.[19,20] After the recombinant plasmid is grown and amplified in an appropriate bacterial host, and then purified, the plasmid template is linearized with a restriction enzyme that cleaves distal to the promoter and adjacent cDNA insert. An appropriate DNA-dependent RNA polymerase is then used to repeatedly transcribe the cloned sequence (in the presence of radiolabeled nucleotide) to yield the labeled probe (Figure 2). A variety of vectors, such as the pSPT18 and pSPT19 plasmids (Boehringer Mannheim) are designed to have the multiple cloning site flanked on either side by different promoters (e.g., Sp6 and T7). With these plasmids, any DNA cloned into the polylinker is transcribed in one direction with Sp6 RNA polymerase and in the opposite direction with T7 RNA polymerase, yielding labeled probes which will be complementary ("antisense") or identical ("sense", as a control) to the mRNA target.

2. Advantages and Disadvantages of cRNA Probes

cRNA probes have several telling advantages over cDNA probes: (1) they are single-stranded, thus avoiding the reannealing problem; (2) they hybridize with greater stability to mRNA, enabling more stringent posthybridization washes; (3) unhybridized probe can be destroyed by posthybridization treatment with RNase which spares the cRNA-mRNA hybrids; and (4) probes of uniform length can be obtained. Accordingly, it has been reported that cRNA probes hybridize significantly better *in situ* than cDNA probes,[16,21] and have thus begun to see more widespread use.[22-39] (For methodological details on the use of these probes, see References 34 to 36.)

Despite the abundant advantages of cRNA probes, they still require some molecular biological expertise to obtain (e.g., subcloning a cDNA fragment into an SP6/T7 promoter-bearing plasmid, followed by growing and amplifying the

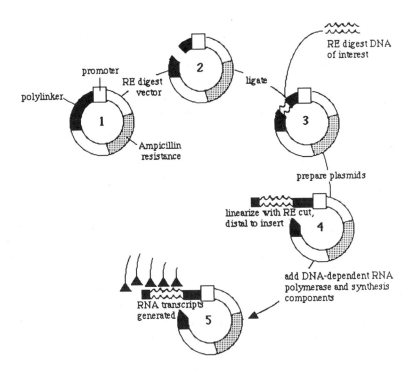

FIGURE 2. Complementary RNA probe labeling using cDNA inserted into specially constructed vector as template for transcription (see text for further explanation).

plasmid in an appropriate bacterial host, etc.), are sensitive to RNases, and may require alkaline hydrolysis into smaller fragments for effective tissue penetration.[21] A further difficulty, which applies to both cDNA and cRNA probes, is that a number of mRNA species are now known to have portions with similar sequences (e.g., the insulin-like growth factors[40-42]). The use of a cloned probe complementary to these homologous regions could readily lead to ambiguous results, a particular difficulty which can be avoided by using probes designed to be uniquely complementary to the nonhomologous regions of mRNAs from a given gene family. Thus, synthetic oligonucleotide probes, as discussed in the following section, may be a useful alternative for some investigators.

C. Synthetic Oligonucleotide Probes
1. Probe Design, Synthesis, and Labeling

The design of synthetic oligonucleotide probes has been discussed elsewhere,[15] and will not be repeated in detail. If the target mRNA sequence is known, probe design is straightforward. The published sequence is generally written 5′ to 3′ (left to right, e.g., 5′-GTCA-3′), so the probe sequence will be complementary from 3′ to 5′ (e.g., 3′-CAGT-5′) although written in the reverse order (e.g., 5′-TGAC-3′). Optimal probe length has not been determined exactly, but 30 to 50 base sequences should form thermally stable hybrids and, in practice, usually give excellent results. The percent G+C content is also relevant since low content (less than 45% G+C) will tend to reduce the thermal stability of the hybrid, while very high content (greater than 65% G+C) may lead to elevated background labeling of the tissue. The selected mRNA region should be compared to other known nucleotide sequences (via commercial DNA database services) to ensure as far as possible that the probe is uniquely complementary to the target mRNA. If several different animal species are being investigated, and the mRNA sequences are known for each species, regions of perfect sequence homology should be utilized in order to obtain a uniformly efficacious probe.

However, if only the amino acid sequence is known, probe design is greatly complicated by codon degeneracy, i.e., the fact that amino acids are generally encoded by more than one RNA base triplet. Consequently, reverse translation of the amino acid sequence into a corresponding nucleic acid results in a set of several possible coding sequences rather than one unique sequence. Instead of synthesizing a mixture of probes reflecting all codon combinations, investigators have devised multiple strategies for designing what should be an optimal probe.[43] Selecting stretches of amino acids with minimal codon degeneracy (particularly methionine and tryptophan, which are uniquely coded), together with the use of species-dependent codon utilization data, are fundamental strategies. Deoxyinosine can be used to replace other deoxynucleotides at several ambiguous sites (e.g., at A/T or G/T ambiguities) in a probe sequence to enhance duplex stability,[44] and thus should be considered in the design of an optimal probe for an ambiguous target sequence. A hypothetical example of the application of some of these strategies is given elsewhere,[15] but their use to date has been limited to the design of oligonucleotide probes to detect target coding sequences in libraries of cloned DNA segments. Computer programs, such as PROBE (Intelligenetics, Inc.), which incorporate known oligonucleotide design strategies, should facilitate the application of this approach to developing probes for *in situ* hybridization histochemistry. In the event that multiple "optimal" probes are used, *in situ* hybridization might be useful in screening for the correct sequence since hybridization conditions could be adjusted to prevent hybridization of a probe with even a single base mismatch.[45,46] The synthesis and purification of synthetic oligonucleotides has been discussed elsewhere,[15,47,48] and suffice to note here that many research institutions, universities, and commercial organizations now have suitable facilities and trained personnel to carry out a custom synthesis and purification for a reasonable fee.

After the oligonucleotide probe is obtained, several labeling options are available: 5′ end-labeling, primer extension, and 3′ end-labeling (Figure 3).

The first method, 5′ end-labeling, uses the enzyme T4 polynucleotide kinase to transfer the terminal phosphate from $[\gamma-{}^{32}P]ATP$ to the free 5′ hydroxyl group of the oligonucleotide. Although the specific activity is limited by the addition

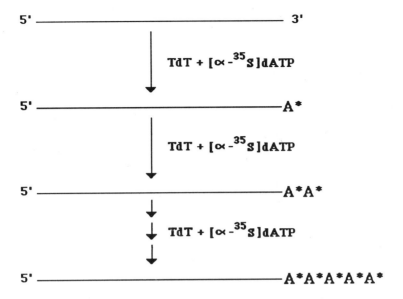

FIGURE 3. Synthetic oligonucleotide probe labeling by progressive enzymatic addition of labeled nucleotide to the 3′ end (see text for further explanation).

of a maximum of one label per molecule of probe, this method has been found to be suitable in some studies of relatively high abundance mRNAs.[15,49-51]

The second method, primer extension labeling,[52] uses the Klenow fragment of DNA polymerase I to catalyze the synthesis of the probe (by extension of a synthetic primer oligonucleotide in the presence of labeled deoxynucleoside triphosphates) across a complementary oligonucleotide template. Although this method has been used to create relatively high specific activity probes for *in situ* hybridization studies,[53-56] the procedure is complicated by the need to prepare both template and primer oligonucleotides, and then separate them electrophoretically to isolate the extended, labeled primer from the reaction mixture. Nevertheless, in contrast to the 5′ end phosphorylation method, lower energy radioisotopic labels (from ^3H, ^{35}S, or ^{125}I-labeled deoxynucleoside triphosphates) can be incorporated into the probe to facilitate high resolution anatomical studies.

The third procedure, 3′ end-labeling, uses the enzyme terminal deoxynucleotidyl transferase to catalyze the sequential addition of radioactive deoxynucleoside monophosphates (from appropriately labeled deoxynucleoside triphosphates) to the free 3′ hydroxyl end of the synthetic oligonucleotide.[57] Since this enzyme will continue adding deoxynucleoside monophosphates to the 3′ end, the specific activity of the probe (i.e., probe length) can be controlled by reaction conditions such as time and substrate concentration. This labeling method has been used for the *in situ* detection of mRNAs encoding proopiomela-nocortin,[15,58] vasopressin,[55,59-62] oxytocin,[60] somatostatin,[62-65] enkephalin,[25,47,66] dynorphin,[66,67] substance P,[66] neuropeptide Y,[68] corticotropin releasing factor,[69,70] vasoactive intestinal polypeptide,[62,71] cholecystokinin,[72] β-amyloid,[73] calmodulin,[74] and tyrosine hydroxylase,[66,75] among others. Collins and Hunsaker[76] have also used 3′ end-labeling to prepare high specific activity probes for genomic blotting studies. The obviously noncomplementary "tail" does not appear to impair either the stability or the specificity of the hybrids.[15,58,70] Electrophoretic separation of probes with different length "tails" is therefore unnecessary, and the labeled probe can be separated from the reaction mixture by a rapid and simple chromatographic step.[55] Thus, while the primer extension and 3′ end-labeling methods share the advantage of incorporating multiple-labeled nucleotides into each molecule of probe, the latter procedure is technically much easier to perform.

2. Advantages and Disadvantages of Oligonucleotide Probes

In contrast to cDNA and cRNA probes, synthetic oligonucleotides do not require any molecular biological expertise to obtain or label. These probes are designed to be complementary to a known mRNA sequence, and are then chemically synthesized by automated apparatus, and labeled enzymatically (see References 15, 34, and 48 for review). Following their successful use as hybridization probes for isolated mRNA in Northern blots,[77-82] synthetic oligonucleotides were successfully used for the *in situ* detection of many mRNAs, as noted above. These numerous studies indicate that synthetic oligonucleotides are useful probes for the detection of a wide variety of target mRNAs *in situ,* and are therefore a viable alternative for the investigator who prefers to avoid the requirements for obtaining biologically derived cDNA or cRNA probes. Nevertheless, since synthetic oligonucleotides generally cannot be labeled to the specific activity which is readily possible with cRNA probes, the study of very rare mRNAs may be more effectively carried out with cRNA probes which share the advantage of being single-stranded and have the further advantages of increased hybrid stability, higher specific activity, and posthybridization enzymatic removal of unhybridized probe.

III. PROBE LABELING OPTIONS

A. Radioisotopic Labels

Probe labeling reactions have been carried out using substrates incorporating ^{3}H, ^{32}P, ^{35}S, and ^{125}I. Although some investigators have succeeded in obtaining cellular resolution with ^{32}P-labeled probes,[25,51] most high-resolution studies have utilized probes labeled with the other three of the above radioisotopes. While tritium offers the highest resolution possible, prolonged exposure times are necessary. ^{125}I-labeled probes permit particularly short exposure times together with high resolution.[55,61,63,84] While high background labeling may sometimes occur with radioiodinated probes,[68] methods for reducing such background have been described.[85] Most investigators have used ^{35}S-labeled probes, which require longer exposure, but generally give low background signals if high concentrations (e.g., 100 mM) of dithiothreitol or β-mercaptoethanol are included in the hybridization buffer.[86]

B. Nonradioisotopic Labels

In recent years several nonradioactive markers have been developed to detect specific nucleotide sequences under a variety of hybridization conditions. These nonradioactive probes are particularly useful for *in situ* hybridization studies where they overcome several limitations normally associated with the use of radiolabeled sequences as probes. For example, hybrids can be detected after only a short period of time, usually less than 24 h with most nonradioactive systems. This significantly reduces the prolonged exposures required for autoradiographic detection of probes labeled with low energy β-emitters (e.g., ^{35}S and ^{3}H). Furthermore, the degree of cellular resolution, a prime requisite for *in situ* hybridization, is significantly improved over that normally achieved with autoradiography. Lastly, the use of these nonradioactive probes avoids the biohazards and additional expenses normally associated with the use of radioisotopes. The following is a brief description of several nonradioactive alternatives to radiolabeled probes that have been successfully used by several laboratories. A more extensive discussion of these probes and their detection is provided in another chapter.

1. Biotinylated Probes

The exceptionally high affinity (10^{-15} M) of the avidin-biotin interaction has formed the basis for a very sensitive method of immunohistochemical detection.[87] One particular advantage of this technology is that several derivatized forms of avidin containing a variety of different reporter groups (e.g., fluorescent groups, colloidal gold, or histochemically detectable enzymes) are recognized by native biotin or biotin complexed to a variety of macromolecules, including proteins and nucleic acids.

Complexing biotin with nucleotides has been particularly useful for hybridization studies. Biotin can be complexed with nucleotides by several methods. Direct labeling can be achieved with either a photoactivatable analog of biotin called photobiotin[88,89] or biotin hydrazide.[90] Alternatively, biotinylated nucleotides (e.g., biotin-11-dUTP or biotin-21-dUTP) can be incorporated enzymatically into a cDNA probe by nick translation[91] or mixed primer extension,[92] or into a synthetic oligonucleotide probe by 3′ end tailing[93] or primer extension.[94] Although each of these labeling methods produces biotinylated DNA probes, the sensitivity of detection has been shown to depend upon the position of the biotin within the probe sequence, with incorporation near the ends favoring higher sensitivity.[95]

Detection of biotinylated probes with various derivitized forms of avidin has been achieved by several laboratories. The most popular method is the use of streptavidin complexed with either alkaline phosphatase or horseradish peroxidase. Both enzymes have been successfully used in *in situ* hybridization in a variety of tissues.[96-99] Antibiotin IgGs can also be used to detect biotinylated nucleotides. Although the affinity of the antibiotin-biotin interaction is considerably less than avidin-biotin, complexing these antibodies with ferritin or colloidal gold produced a sufficiently high degree of resolution for electron microscopy.[100] Presumably, similar results could be achieved with the use of avidin complexed to ferritin or colloidal gold.

2. Enzyme-Conjugated Probes

Several different enzymes have been successfully coupled to nucleotides. Histochemical studies have focused on the use of alkaline phosphatase or horseradish peroxidase for the detection of hybridization signals. Ruth[101] has described a method for chemically incorporating modified bases with functionalized "linker arms" into synthetic oligonucleotides. These linker arms can be conjugated to several different enzymes. This approach offers precise control over the location and number of labeling sites within the oligonucleotide, and affords a novel opportunity for the direct nonradioactive stochiometric detection of relatively rare mRNAs *in situ*.[102] Moreover, the sensitivity of these alkaline phosphatase-conjugated probes reportedly exceeds that which can be obtained with biotinylated probes.[103] An alternative method for coupling these enzymes to single-stranded DNA probes was reported by Renz and Kurz,[104] who converted the enzymes into DNA binding proteins by chemical conjugation with polyethyleneimine.

3. Hapten-Conjugated Probes

The hapten-conjugation approach to probe detection was pioneered by Landegent et al.[105] and Tchen et al.,[106] who found that guanine residues in nucleic acids could be modified by treatment with *N*-acetoxy-*N*-2-acetylaminofluorene and then recognized with specific antibodies. This approach has been applied to the detection of single genes in chromosomes,[107] but apparently not yet to mRNA in tissue sections.

Recently, a novel *in situ* hybridization method has been developed based on the enzymatic incorporation of digoxigenin-dUTP (Boehringer Mannheim Biochemicals) into synthetic oligonucleotide probes.[108] The digoxigenin-dUTP labeled probe is hybridized and then detected with an alkaline phosphatase-conjugated IgG which is highly specific for the digoxigenin molecule (Figure 4). Unlike the direct conjugation of a single molecule of alkaline phosphatase to each molecule of probe,[101] this new methodology offers the advantage of incorporating multiple labels into the probe, thus amplifying the final hybridization signal. In addition, IgG detection also affords the advantage of using antibody "bridging" techniques which are known to result in considerable signal amplification in immunocytochemistry.

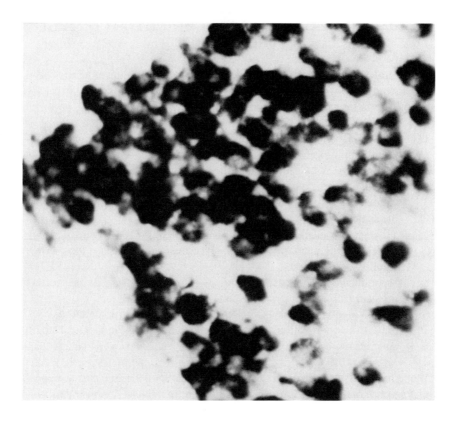

FIGURE 4. Enzyme histochemical detection of vasopressin mRNA in rat supraoptic nucleus, using synthetic oligonucleotide probe labeled on the 3′ end (see Figure 3) with digoxigenin-dUTP (Boehringer Mannheim Biochemicals).

4. Fluorescent Probes

Although fluorescent markers appear to be less sensitive than enzyme-based detection systems,[109,110] they have the advantage of direct detection, as well as the potential for the simultaneous detection of multiple probes labeled with spectrally discriminable fluorophores. Fluorophores may be introduced into probes chemically or enzymatically (see Reference 109 for other references) and have been successfully used in several laboratories for *in situ* hybridization[111] or the detection of immobilized genomic DNA or RNA. Although these procedures are generally less sensitive than those employing enzyme-linked or radiolabeled probes, they are nonetheless valuable for certain experimental and clinical applications (e.g., detection of viral infections), and have been reported to provide superior resolution.[112]

IV. IMPLICATIONS OF PROBE CHOICE FOR CONTROL PROCEDURES

Control procedures to ensure the specificity of *in situ* hybridization have been considered in more detail elsewhere[15,34,36,58,86,113] and are mentioned only briefly here. Criteria for hybridization specificity include:

1. Colocalization of the hybridization signal with appropriate immunoreactive staining for the encoded precursor
2. Competition of hybridization of the labeled probe by unlabeled probe of the same sequence
3. Colocalization of hybrids formed by different probes complementary to the same mRNA target
4. Thermal dissociation of the hybrid at a temperature (Tm_{obs}) consistent with perfect complementarity (Tm_{calc}), since mismatches reduce the thermal stability of the hybrid
5. Hybridization of the labeled probe to an RNA target of the correct molecular size (Northern analysis)
6. Pretreatment of the tissue with RNase to destroy the mRNA target
7. Failure of hybridization by a noncomplementary probe of identical length and G/C composition

While the relative merits of these various controls are debatable, it is also clear that no one control procedure alone is sufficient to reliably demonstrate specificity of probe hybridization *in situ*. Given this caveat, it is still worth noting that probe choice has some bearing on which control procedures may be carried out by the investigator.

With both cDNA and cRNA probes, it may be impossible to carry out probe colocalization studies unless two nonoverlapping cloned sequences are obtained. Futhermore, thermal dissociation (Tm) studies with nick-translated cDNA probes are virtually meaningless since a wide variety of labeled fragments of varying lengths and G-C content are present (see Figure 1). Such studies may also be difficult with cRNA probes unless care is taken to separate full length probe from incomplete or partly degraded probe following the labeling reaction.[36] In addition, the RNase pretreatment control may not be appropriate for cRNA probes since residual enzyme could digest the probe itself;[34] however, this artifact may be prevented by digestion of the RNase with proteinase K prior to hybridization.[36] None of these limitations pertain to synthetic oligonucleotide probes, for which the full range of control studies can and should be carried out to ensure hybridization specificity.

V. CONCLUSIONS

A. Issues Determining Probe and Label Choice
The choice of probe and label depends upon the resources available to the

experimenter as well as the requirements of the experiment. In terms of resources, if the experimenter does not have access to the required clone, or does not have access to the relevant molecular biological and microbiological techniques, choice of probe will be restricted to synthetic oligonucleotides. If these limitations do not apply, the abundance of the target mRNA may be a deciding factor; the greater possible specific activity of cRNA compared to synthetic oligonucleotide probes may mandate their use for the detection of very rare target mRNAs. Since cRNA probes have numerous advantages over cDNA probes, but share the need for clone access and molecular biological/microbiological techniques, most investigators using cDNAs may wish to subclone their DNA inserts into appropriate plasmids containing RNA polymerase promoters. While the cDNA probes will certainly be usable for many relatively abundant mRNA targets, the greater speed of detection with cRNA probes (e.g., 3 d vs. 3 week autoradiographic exposures) will promote their use even where detection limits are not at issue.

The need for resolution may constrain the type of label used for any given type of hybridization probe. For example, in a tissue where expected positive and negative cells are anatomically adjacent, the loss of resolution due to the use of a high energy radioisotope may preclude interpretation of the data. Resolution will clearly be enhanced by the use of tritium- or enzyme-labeled probes. Unfortunately, these two classes of probes also provide information which is intrinsically difficult to quantify. Tritium-labeled probes emit low-energy β-particles which are absorbed by the tissue in relation to the distance of the emitter from the photographic emulsion, and enzyme-labeled probes are detected by the formation of an insoluble reaction product which bears a complex, nonlinear relation to the amount of enzyme (and thus probe) which is present. Thus, the need for resolution and quantitation are somewhat in conflict, and the investigator must determine which is most critical for any particular application.

B. Future Prospects for Hybridization Probes

In situ hybridization histochemistry has yielded valuable insights into the distribution and regulation of mRNAs within cellular elements in a variety of tissue preparations. Of particular note is its value in neural tissue where the resolution of mRNA in single cells is mandatory. An exciting extension of this technology has been developed in the use of intervening sequence probes to detect rapid alterations in primary transcripts within individual cell nuclei, prior to processing and transport of the mature mRNA to the cytoplasm.[114] The use of these probes can help in interpreting changes in mRNA levels, detected *in situ,* as being caused by changes in gene expression rather than alterations in mRNA stability or degradation rate.

Synthetic oligonucleotide probes appear to be particularly useful in the study of gene expression. For example, small oligonucleotide probes complementary to specific exon coding regions may be used to study alternate RNA processing at the level of single cells. Similarly, intervening sequence probes may be designed to assess gene transcription in cellular nuclei. Thus, because of the ease of the design and synthesis of oligonucleotide probes, they have much potential in the cellular analysis of RNA transcription and processing. Perhaps the full potential of synthetic oligonucleotides for *in situ* hybridization will be realized by the "cloning" of these probes with cRNA synthesizing vectors,[115] resulting in the generation of high specific activity probes which share the advantage of increased thermal stability with "true" cRNA probes prepared as described above. To date, there have been no reports on the use of biotinylated or hapten-coupled cRNA probes for *in situ* hybridization, but these may prove to be particularly advantageous for the nonradioactive detection of very low abundance RNAs.

One significant limitation in detecting mRNA within individual cells *in situ* is the low copy number for many species of mRNA (e.g., neuroreceptor proteins in the nervous system). A recent study has described a novel technology termed *in situ* transcription[116] which may ultimately alleviate this problem. In this method an oligonucleotide is used to prime the reverse transcription of DNA from an mRNA template *in situ*. These cDNAs can contain many radioactive nucleotides and thus have the potential to permit the detection of relatively rare mRNAs in single cells. However, the real power of this new technology may be the cloning of cDNAs directly from tissue sections. It may be possible to further amplify *in situ* transcription signals by making use of the polymerase chain reaction (PCR) in which two primers are used in the exponential enzymatic amplification of DNA[117] or mRNA.[118] Although this technology has been limited to the detection of DNA or mRNA *in vitro*, it may be feasible to carry out PCR amplification *in situ*, which would then greatly increase the hybridization signal from very rare mRNAs within single cells. While this method would be intrinsically nonquantitative, it would have the virtue of amplifying the probe target and thus extending the effective sensitivity of any nucleic acid probe.

ACKNOWLEDGMENTS

We wish to acknowledge the superb efforts of Ms. Elaine Robbins and Ms. Kathy Callison in developing the nonradioactive, antibody-based *in situ* hybridization method mentioned here, as well as our colleagues at Boehringer Mannheim, Dr. Debra Grega and Dr. Richard L. Martin, for their cooperation in its development.

REFERENCES

1. **Gall, J. and Pardue, M.,** The formation and detection of RNA-DNA hybrid molecules in cytological preparations, *Proc. Natl. Acad. Sci. U.S.A.,* 63, 378 1969.
2. **John, H., Birnstiel, M., and Jones, K.,** RNA-DNA hybrids at the cytological level, *Nature,* 223, 582, 1969.

3. **Gall, J. and Pardue, M.,** Nucleic acid hybridization in cytological preparations, *Methods Enzymol.,* 21, 470, 1971.

4. **Wimber, D. E. and Steffensen,** Localization of gene function, *Ann. Rev. Biochem.,* 20, 205, 1973.

5. **Wiener, F., Spira, J., Banerjee, M., and Klein, G.,** A new approach to gene mapping by *in situ* hybridization on isolated chromosomes, *Som. Cell Mol. Gen.,* 11, 493, 1985.

6. **Harrison, P. R., Conkie, D., Paul, J., and Jones, K.,** Localization of cellular globin messenger RNA by *in situ* hybridization to complementary DNA, *FEBS Lett.,* 32, 109, 1973.

7. **Pochet, R., Brocas, H., Vassart, G., Toubeau, G., Seo, H., Refetoff, S., Dumont, J. E., and Pasteels, J. L.,** Radioautographic localization of prolactin messenger RNA on histological sections by *in situ* hybridization, *Brain Res.,* 211, 433, 1981.

8. **Ausubel, F. M., Brent, R., Kingston, R. E., Moore, D. D., Seidman, J. G., Smith, J. A., and Struhl, K., Eds.,** *Current Protocols in Molecular Biology,* John Wiley & Sons, New York, 1987.

9. **Rigby, P. W. J., Dieckmann, M., Rhodes, C., and Berg, P.,** Labeling deoxyribonucleic acid to high specific activity *in vitro* by nick translation with DNA polymerase I, *J. Mol. Biol.,* 113, 237, 1977.

10. **Feinberg, A. P. and Vogelstein, B.,** A technique for radiolabeling DNA restriction endonuclease fragments to high specific activity, *Anal. Biochem.,* 132, 6, 1983.

11. **Feinberg, A. P. and Vogelstein, B.,** Addendum: a technique for radiolabeling DNA restriction endonuclease fragments to high specific activity, *Anal. Biochem.,* 137, 266, 1984.

12. **Gee, C. E. and Roberts, J. L.,** *In situ* hybridization histochemistry: a technique for the study of gene expression in single cells, *DNA,* 2, 157, 1983.

13. **Shivers, B. D., Schacter, B. S., and Pfaff, D. W.,** *In situ* hybridization for the study of gene expression in the brain, *Methods Enzymol.,* 124, 497, 1986.

14. **Wilcox, J. N., Gee, C. E., and Roberts, J. L.,** *In situ* cDNA:mRNA hybridization: development of a technique to measure mRNA levels in individual cells, *Methods Enzymol.,* 124, 510, 1986.

15. **Lewis, M. E., Sherman, T. G., and Watson, S. J.,** *In situ* hybridization histochemistry with synthetic oligonucleotides: strategies and methods, *Peptides,* 6(Suppl. 2), 75, 1985.

16. **Angerer, R. C., Cox, K. H., and Angerer, L. M.,** *In situ* hybridization to cellular RNAs, in *Genetic Engineering,* Vol. 7, Setlow, J. K. and Hollaender, A., Eds., Plenum Press, New York, 1985, 43.

17. **Hu, N. T. and Messing, J.,** The making of strand-specific M13 probes, *Gene,* 17, 271, 1982.

18. **Lama, E., Kahn, A., and Guillouzo, A.,** Detection of mRNAs present at low concentration in rat liver by *in situ* hybridization: application to the study of metabolic regulation and azo dye hepatocarcinogenesis, *J. Histochem. Cytochem.,* 35, 559, 1987.

19. **Johnson, M. T. and Johnson, B. A.,** Efficient synthesis of high specific activity ^{35}S-labelled human beta-globin pre-mRNA, *BioTechniques,* 2, 156, 1984.

20. **Melton, D. A., Krieg, P. A., Rebagliati, M. R., Maniatis, T., Zinn, K., and Green, M. R.,** Efficient *in vitro* synthesis of biologically active RNA and RNA hybridization probes from plasmids containing a bacteriophage SP6 promoter, *Nucl. Acids Res.,* 12, 7035, 1984.

21. **Cox, K. H., DeLeon, D. V., Angerer, L. M., and Angerer, R. C.,** Detection of mRNAs in sea urchin embryos by *in situ* hybridization using asymmetric RNA probes, *Dev. Biol.,* 101, 485, 1984.

22. **Lynn, D. A., Angerer, L. M., Bruskin, A. M., Klein, W. H., and Angerer, R. C.,** Localization of a family of mRNAs in a single cell type and its precursors in sea urchin embryos, *Proc Natl. Acad. Sci. U.S.A.,* 80, 2656, 1983.

23. **Knipple, D. C., Seifert, E., Rosenberg, U. B., Preiss, A., and Jackle, H.,** Spatial and temporal patterns of Kruppel gene expression in early *Drosophila* embryos, *Nature,* 317, 40, 1985.

24. **Siegel, R. E. and Young, W. S., III,** Detection of preprocholecystokinin and preproenkephalin A mRNAs in rat brain by hybridization histochemistry using complementary RNA probes, *Neuropeptides,* 6, 573, 1985.

25. **Bloch, B., Popovici, T., LeGuellec, D., Normand, E., Chouham, S., Guitteny, A. F., and Bohlen, P.,** *In situ* hybridization histochemistry for the analysis of gene expression in the endocrine and central nervous system tissues: a three year experience, *J. Neurosci. Res.,* 16:183, 1986.

26. **Goldman, D., Simmons, D., Swanson, L. W., Patrick, J., and Heinemann, S.,** Mapping of brain areas expressing RNA homologous to two different acetylcholine receptor alpha-subunit cDNAs, *Proc. Natl. Acad. Sci. U.S.A.,* 83, 4076, 1986.

27. **Hoefler, H., Childers, H., Montminy, M. R., Lechan, R. M., Goodman, R. H., and Wolfe, H. J.,** *In situ* hybridization methods for the detection of somatostatin mRNA in tissue sections using antisense RNA probes, *Histochem. J.,* 18, 597, 1986.

28. **Allen, J. M., Sasek, C. A., Martin, J. B., and Heinrich, G.,** Use of complementary ^{125}I-labelled RNA for single cell resolution by *in situ* hybridization, *BioTechniques,* 5, 774, 1987.

29. **Bahmanyar, S., Higgins, G. A., Goldgaber, G., Lewis, D. A., Morrison, J. H., Wilson, M. C., Shankar, S. K., and Gajdusek, D. C.,** Localization of amyloid beta protein messenger RNA in brains from patients with Alzheimer's disease, *Science,* 237, 77, 1987.

30. **Chesselet, M.-F. and Affolter, H. U.,** Preprotachykinin messenger RNA detected by *in situ* hybridization in striatal neurons of the human brain, *Brain Res.,* 410, 83, 1987.

31. **Chesselet, M.-F., Weiss, L., Wuenschell, C., Tobin, A. J., and Affolter, H. U.,** Comparative distribution of mRNA's for tachykinins in the basal ganglia: an *in situ* hybridization study in the rodent brain, *J. Comp. Neurol.,* 262, 125, 1987.

32. **Terenghi, G., Polak, J. M., Hamid, Q., O'Brien, E., Denny, P., Legon, S., Dixon, J., Minth, C. D., Palay, S. L., and Yasargil, G.,** Localization of neuropeptide Y mRNAs in neurons of human cerebral cortex by means of *in situ* hybridization with a complementary RNA probe, *Proc. Natl. Acad. Sci. U.S.A.,* 84, 7315, 1987.

33. **Watson, S. J., Sherman, T. G., Kelsey, J. E., Burke, S., and Akil, H.,** Anatomical localization of mRNA: *in situ* hybridization of neuropeptide systems, in *In Situ Hybridization: application to Neurobiology,* Valentino, K., Eberwine, J., and Barchas, J., Eds., Oxford University Press, New York, 1987, 126.

34. **Baldino, F., Jr., Chesselet, M.-F., and Lewis, M. E.,** High resolution *in situ* hybridization histochemistry, *Methods Enzymol.,* 168, 79, 1989.

35. **Palmert, M. R., Golde, T. E., Cohen, M. L., Kovacs, D. M., Tanzi, R. E., Gusella, J. F., Usiak, M. F., Younkin, L. H., and Younkin, S. G.,** Amyloid protein precursor messenger RNAs: differential expression in Alzheimer's disease, *Science,* 241, 1080, 1988.

36. **Watson, S. J., Patel, P., Burke, S., Herman, J., Schafer, M., and Kwak, S.,** *In situ* hybridization of mRNA in nervous tissue: a primer, in *In Situ Hybridization and Related Techniques to Study Cell-Specific Gene Expression in the Nervous System,* Roberts, J. L., Ed., Society for Neuroscience, Washington, D.C., 1988, 4.

37. **Chesselet, M. F., Weiss, L. T., and Robbins, E.,** Quantitative variations of messenger RNA levels in neurons of the basal ganglia: an *in situ* hybridization study, in *Brain Imaging: Techniques and Applications,* Sharif, N. A. and Lewis, M. E., Eds., Ellis Horwood, Ltd., Chichester, U.K., 1989, 220.

38. **Schafer, M. K.-H., Day, R., Herman, J. P., Kwasiborski, V., Sladek, C. D., Akil, H., and Watson, S. J.,** Effects of electroconvulsive shock on dynorphin in the hypothalamic-neurohypophyseal system of the rat, in *Advances in the Biosciences, International Narcotics Research Conference,* Hamon, M., Ed., Albi, France, in press.

39. **Wada, E., Wada, K., Boulter, J., Deneris, E., Heinemann, S., Patrick, J., and Swanson, L. W.,** The distribution of alpha2, alpha3, alpha4, and beta2 neuronal nicotinic receptor subunit mRNAs in the central nervous system. A hybridization histochemical study in the rat, *J. Comp. Neurol.,* in press.

40. **Irminger, J. C., Rosen, K. M., Humbel, R. E., and Villa-Komaroff, L.,** Tissue-specific expression of insulin-like growth factor II mRNAs with distinct 5' untranslated regions, *Proc. Natl. Acad. Sci. U.S.A.,* 84, 6330, 1987.

41. **Roberts, C. T., Jr., Lasky, S. R., Lowe, W. L., Jr., Seaman, W. T., and LeRoith, D.,** Molecular cloning of rat insulin-like growth factor I complementary deoxyribonucleic acids: differential messenger ribonucleic acid processing and regulation by growth hormone in extrahepatic tissues, *Mol. Endocrinol.,* 1, 243, 1987.

42. **Rotwein, P., Burgess, S. K., Milbrandt, J. D., and Krause, J. E.,** Differential expression of insulin-like growth factor genes in rat central nervous system, *Proc. Natl. Acad. Sci. U.S.A.,* 85, 265, 1988.

43. **Lathe, R.,** Synthetic oligonucleotide probes deduced from amino acid sequence data, *J. Mol. Biol.,* 183, 1, 1985.

44. **Martin, F. H., Castro, M. M., Aboul-ela, F., and Tinoco, I., Jr.,** Base pairing involving deoxyinosine: implications for probe design, *Nucl. Acids Res.,* 13, 8927, 1985.

45. **Wallace, R. B., Shaffer, J., Murphy, R. F., Bonner, J., Hirose, T., and Itakura, K.,** Hybridization of synthetic oligodeoxyribonucleotides to 174 DNA: the effect of single base pair mismatch, *Nucl. Acids Res.,* 6, 3543, 1979.

46. **Buvoli, M., Biamonti, G., Riva, S., and Morandi, C.,** Hybridization of oligonucleotide probes to RNA molecules: specificity and stability of duplexes, *Nucl. Acids Res.,* 15, 9091, 1987.

47. **Gait, M. J.,** *Oligonucleotide Synthesis: A Practical Approach,* IRL Press, New York, 1986.

48. **Lewis, M. E., Krause, R. G., II, and Roberts-Lewis, J. M.,** Recent developments in the use of synthetic oligonucleotides for *in situ* hybridization histochemistry, *Synapse,* 2, 308, 1988.

49. **Arentzen, R., Baldino, F., Jr., Davis, L. G., Higgins, G. A., Lin, Y., Manning, R. W., and Wolfson, B.,** *In situ* hybridization of putative somatostatin mRNA within hypothalamus of the rat using synthetic oligonucleotide probes, *J. Cell Biochem.,* 27, 415, 1985.

50. **Nojiri, H., Sato, M., and Urano, A.,** *In situ* hybridization of the vasopressin mRNA in the rat hypothalamus by use of a synthetic oligonucleotide probe, *Neurosci. Lett.,* 58, 101, 1985.

51. **Wolfson, B., Manning, R. W., Davis, L. G., Arentzen, R., and Baldino, F., Jr.,** Co-localization of corticotropin releasing factor and vasopressin mRNA in neurones after adrenalectomy, *Nature,* 315, 59, 1985.

52. **Studencki, A. B. and Wallace, R. B.,** Allele-specific hybridization using oligonucleotide probes of very high specific activity: discrimination of the human beta^A- and beta^S-globin genes, *DNA,* 3, 7, 1984.

53. **Uhl, G. R., Zingg, H. H., and Habener, J. F.,** Vasopressin mRNA *in situ* hybridization: localization and regulation studied with oligonucleotide cDNA probes in normal and Brattleboro rat hypothalamus, *Proc. Natl. Acad. Sci. U.S.A.,* 82, 5555, 1985.

54. **Uhl, G. R. and Sasek, C. A.,** Somatostatin mRNA: regional variation in hybridization densities in individual neurons, *J. Neurosci.,* 6, 3258, 1986.

55. **Lewis, M. E., Arentzen, R., and Baldino, F., Jr.,** Rapid, high resolution *in situ* hybridization histochemistry with radioiodinated synthetic oligonucleotides, *J. Neurosci. Res.,* 16, 117, 1986.

56. **Morris, B. J., Haarmann, I., Kempter, B., Hollt, V., and Herz, A.,** Localization of prodynorphin mRNA in rat brain by *in situ* hybridization using a synthetic oligonucleotide probe, *Neurosci. Lett.,* 69, 104, 1986.

57. **Bollum, F. J.,** Terminal deoxynucleotidyl transferase, in *The Enzymes,* Vol. 10, Boyer, P. D., Ed., Academic Press, New York, 1974, 145.

58. **Lewis, M. E., Sherman, T. G, Burke, S., Akil, H., Davis, L. G., Arentzen, R., and Watson, S. J.,** Detection of proopiomelanocortin mRNA by *in situ* hybridization with an oligonucleotide probe, *Proc. Natl. Acad. Sci. U.S.A.,* 83, 5419, 1986b.

59. **Baldino, F., Jr. and Davis, L. G.,** Glucocorticoid regulation of vasopressin messenger RNA, in *In Situ Hybridization in Brain,* Uhl, G. R., Ed., Plenum Press, New York, 1986, 97.

60. **Young, W. S., III, Mezey, E., and Siegel, R.,** Vasopressin and oxytocin mRNAs in adrenalectomized and Brattleboro rats: analysis by quantitative *in situ* hybridization histochemistry, *Mol. Brain Res.,* 1, 231, 1986.

61. **Baldino, F., Jr., O'Kane, T. M., Fitzpatrick-McElligott, S., and Wolfson, B.,** Coordinate hormonal and synaptic regulation of vasopressin messenger RNA, *Science,* 241, 978, 1988.

62. **Card, J. P., Fitzpatrick-McElligott, S., Gozez, I., and Baldino, F., Jr.,** Localization of vasopressin, vasoactive intestinal polypeptide, peptide histidine-isoleucine, and somastostatin mRNA in rat suprachiasmatic nucleus, *Cell Tissue Res.,* 252, 307, 1988.

63. **Rogers, W. T., Schwaber, J. S., and Lewis, M. E.,** Quantitation of cellular resolution *in situ* hybridization histochemistry in brain by image analysis, *Neurosci. Lett.,* 82, 315, 1987.

64. **Fitzpatrick-McElligott, S., Card, J. P., Lewis, M. E., and Baldino, F., Jr.,** Neuronal localization of prosomatostatin mRNA in rat brain with *in situ* hybridization histochemistry, *J. Comp. Neurol.,* 273, 558, 1988.

65. **Baldino, F., Jr., Fitzpatrick-McElligott, S., O'Kane, T. M., and Gozez, I.,** Hormonal regulation of somatostatin mRNA, *Synapse,* 2, 317, 1988.

66. **Young, W. S., III, Bonner, T. I., and Brann, M. R.,** Mesencephalic dopamine neurons regulate the expression of neuropeptide mRNAs in the rat forebrain, *Proc. Natl. Acad. Sci. U.S.A.,* 83, 9827, 1986.

67. **Ruda, M. A., Iadarola, M. J., Cohen, L. V., and Young, W. S., III,** *In situ* hybridization histochemistry and immunocytochemistry reveal an increase in spinal dynorphin biosynthesis in a rat model of peripheral inflammation and hyperalgesia, *Proc. Natl. Acad. Sci. U.S.A.,* 85, 622, 1988.

68. **Gehlert, D. R., Chronwall, B. M., Schafer, M. P., and O'Donohue, T. L.,** Localization of neuropeptide Y messenger ribonucleic acid in rat and mouse brain by *in situ* hybridization, *Synapse,* 1, 25, 1987.

69. **Young, W. S., Mezey, E., and Siegel, R.,** Quantitative *in situ* hybridization histochemistry reveals increased levels of corticotropin-releasing factor mRNA after adrenalectomy in rats, *Neurosci. Lett.,* 70, 198, 1986.

70. **Palkovits, M., Leranth, C., Gorcs, T., and Young, W. S., III,** Corticotropin-releasing factor in the olivocerebellar tract of rats: demonstration by light- and electron-microscopic immunohistochemistry and *in situ* hybridization histochemistry, *Proc. Natl. Acad. Sci. U.S.A.,* 84, 3911, 1987.

71. **Baldino, F., Jr., Gozez, I., Fitzpatrick-McElligott, S., and Card, J. P.,** Localization of vasoactive polypeptide and PHI-27 messenger RNA in rat thalamic and cortical neurons, *J. Mol. Neurosci.,* 1, 199, 1989.

72. **Ingram, S. M., Krause, R. G., II, Baldino, F., Jr., Skeen, L. C., and Lewis, M. E.,** Neuronal localization of cholecystokinin mRNA in rodent brain with *in situ* hybridization histochemistry, *J. Comp. Neurol.,* 287, 260, 1989.

73. **Cohen, M. L., Golde, T. E., Usiak, M. F., Younkin, L. H., and Younkin, S. G.,** *In situ* hybridization of nucleus basalis neurons shows increased beta-amyloid mRNA in Alzheimer disease, *Proc. Natl. Acad. Sci. U.S.A.,* 85, 1227, 1988.

74. **Roberts-Lewis, J. M., Cimino, M., Krause, R. G., II, Tyrrell, D. F., Jr., Davis, L. G., Weiss, B., and Lewis, M. E.,** Distribution of calmodulin mRNA in rat brain using cloned cDNA and synthetic oligonucleotide probes, *Synapse,* in press.

75. **Baldino, F., Jr., Deutch, A. Y., Roth, R. H., and Lewis, M. E.,** *In situ* hybridization histochemistry of tyrosine hydroxylase mRNA in rat brain, *Ann. N.Y. Acad. Sci.,* 537, 484, 1988.

76. **Collins, M. L. and Hunsacker, W. R.,** Improved hybridization assays employing tailed oligonucleotide probes: a direct comparison with 5′-end-labeled oligonucleotide probes and nick-translated plasmid probes, *Anal. Biochem.,* 151, 211, 1985.

77. **Mevarech, M., Noyes, B. E., and Agarwal, K. L.,** Detection of gastrin-specific mRNA using oligodeoxynucleotide probes of defined sequence, *J. Biol. Chem.,* 254, 7472, 1979.

78. **Noyes, B. E., Mevarech, M., Stein, R., and Agarwal, K.,** Detection and partial sequence analysis of gastrin mRNA by using an oligodeoxynucleotide probe, *Proc. Natl. Acad. Sci. U.S.A.,* 76, 1770, 1979.

79. **Agarwal, K. L., Brunstedt, J., and Noyes, B. E.,** A general method for detection and characterization of an mRNA using an oligonucleotide probe, *J. Biol. Chem.,* 256, 1023, 1981.

80. **Gubler, U., Kilpatrick, D. L., Seeburg, P. H., Gage, L. P., and Udenfriend, S.,** Detection and partial characterization of proenkephalin mRNA, *Proc. Natl. Acad. Sci. U.S.A.,* 78, 5484, 1981.

81. **Comb, M., Herbert, E., and Crea, R.,** Partial characterization of the mRNA that codes for enkephalins in bovine adrenal medulla and human pheochromocytoma, *Proc. Natl. Acad. Sci. U.S.A.,* 79, 360, 1982.

82. **Mazjoub, J. A., Rich, A., van Boom, J., and Habener, J. F.,** Vasopressin and oxytocin mRNA regulation in the rat assessed by hybridization with synthetic oligonucleotides, *J. Biol. Chem.,* 258, 14061, 1983.

83. **Davis, L. G., Arentzen, R., Reid, J., Manning, R. W., Wolfson, B., Lawrence, K. L., and Baldino, F., Jr.,** Glucocorticoid-sensitive vasopressin mRNA synthesis in the paraventricular nucleus of the rat, *Proc. Natl. Acad. Sci. U.S.A.,* 83, 1145, 1986.

84. **Allen, J. M., Sasek, C. A., Martin, J. B., and Heinrich, G.,** Use of complementary ^{125}I-labeled RNA for single cell resolution by *in situ* hybridization, *BioTechniques,* 5, 774, 1987.

85. **Hayashi, S., Gilham, I. C., Delaney, A. D., and Tener, G. M.,** Acetylation of chromosome squashes from *Drosophila melanogaster* decreases the background in autoradiographs from hybridization with ^{125}I-labeled RNA, *J. Histochem. Cytochem.,* 26, 677, 1978.

86. **Young, W. S., III,** *In situ* hybridization histochemistry, in *Handbook of Chemical Neuroanatomy,* Vol. 8, *Neuronal Microcircuits,* Bjorklund, A. and Hokfelt, T., Eds., Elsevier, New York, in press.

87. **Hsu, S.-M., Raine, L., and Fanger, H.,** Use of avidin-biotin-peroxidase complex (ABC) in immunoperoxidase techniques: a comparison between ABC and unlabeled antibody (PAP) procedures, *J. Histochem. Cytochem.,* 29, 577, 1981.

88. **Forster, A. C., McInnes, J. L., Skingle, J. C., and Symons, R. H.,** Non-radioactive hybridization probes prepared by the chemical labeling of DNA and RNA with a novel reagent, photobiotin, *Nucl. Acids Res.,* 13, 745, 1985.

89. **McInnes, J. L., Dalton, S., Vize, P. D., and Robins, A. J.,** Non-radioactive photobiotin-labeled probes detect single copy genes and low abundance mRNA, *Bio/Technology,* 5, 269, 1987.

90. **Reisfeld, A., Rothenberg, J. M., Bayer, E. A., and Wilchek, M.,** Nonradioactive hybridization probes prepared by the reaction of biotin hydrazide with DNA, *Biochem. Biophys. Res. Commun.,* 142, 519, 1987.

91. **Langer, P. R., Waldrop, A. A., and Ward, D. C.,** Enzymatic synthesis of biotin-labeled polynucleotides: novel nucleic acid affinity probes, *Proc. Natl. Acad. Sci. U.S.A.,* 78, 6633, 1981.

92. **Kincaid, R. L. and Nightingale, M. S.,** A rapid nonradioactive procedure for plaque hybridization using biotinylated probes prepared by random primed labeling, *BioTechniques,* 6, 44, 1988.

93. **Rigas, B., Welcher, A. A., Ward, D. C., and Weissman, S. M.,** Rapid plasmid library screening using RecA-coated biotinylated probes, *Proc. Natl. Acad. Sci. U.S.A.,* 83, 9591, 1986.

94. **Murasugi, A. and Wallace, R. B.,** Biotin labeled oligonucleotides: enzymatic synthesis and use as hybridization probes, *DNA,* 3, 269, 1984.

95. **Cook, A. F., Vuocolo, E., and Brakel, C. L.,** Synthesis and hybridization of a series of biotinylated oligonucleotides, *Nucl. Acids Res.,* 16, 4077, 1988.

96. **Eng, L. F., Stocklin, E., Lee, Y.-L., Shiurba, R. A., Coria, F., Halks-Miller, M., Mozsgai, C., Fukayama, G., and Gibbs, M.,** Astrocyte culture on nitrocellulose membranes and plastic: detection of cytoskeletal proteins and mRNAs by immunocytochemistry and *in situ* hybridization, *J. Neurosci. Res.,* 16, 239, 1986.

97. **Unger, E. R., Budgeon, L. R., Myerson, D., and Brigati, D. J.,** Viral diagnosis by *in situ* hybridization. Description of a rapid simplified colorimetric method, *Am. J. Surg. Pathol.,* 10, 1, 1986.

98. **Arai, H., Emson, P. C., Agrawal, S., Christodoulou, C., and Gait, M. J.,** *In situ* hybridization histochemistry: localization of vasopressin mRNA in rat brain using a biotinylated oligonucleotide probe, *Mol. Brain Res.,* 4, 63, 1988.

99. **Guitteny, A.-F., Fouque, B., Mougin, C., Teoule, R., and Bloch, B.,** Histological detection of messenger RNAs with biotinylated synthetic oligonucleotide probes, *J. Histochem. Cytochem.,* 36, 563, 1988.

100. **Binder, M., Tourmente, S., Roth, J., Renaud, M., and Gehring, W. J.,** *In situ* hybridization at the electron microscope level: localization of transcripts on ultrathin sections of Lowicryl K4M-embedded tissue using biotinylated probes and protein A-gold complexes, *J. Cell Biol.,* 102, 1646, 1986.

101. **Ruth, J. L.,** Chemical synthesis of non-radioactively labeled DNA hybridization probes, *DNA,* 3, 123, 1984.

102. **Baldino, F., Jr., Ruth, J. L., and Davis, L. G.,** Non-radioactive detection of vasopressin mRNA with *in situ* hybridization histochemistry, *Exp. Neurol.,* 104, 200, 1989.

103. **Jablonski, E., Moomaw, E. W., Tullis, R. H., and Ruth, J. L.,** Preparation of oligode-oxynucleotide-alkaline phosphatase conjugates and their use as hybridization probes, *Nucl. Acids Res.,* 14, 6115, 1986.

104. **Renz, M. and Kurz, C.,** A colorimetric method for DNA hybridization, *Nucl. Acids Res.,* 12, 3435, 1984.

105. **Landegent, J. E., Jansen in de Wal, N., Baan, R. A., Hoeijmakers, J. H. J., and Van der Ploeg, M.,** 2-Acetylaminofluorene-modified probes for the indirect hybridocytochemical detection of specific nucleic acid sequences, *Exp. Cell Res.,* 153, 61, 1984.

106. **Tchen, P., Fuchs, R. P. P., Sage, E., and Leng, M.,** Chemically modified nucleic acids as immunodetectable probes in hybridization experiments, *Proc. Natl. Acad. Sci. U.S.A.,* 81, 3466, 1984.

107. **Landegent, J. E., Jansen in de Wal, N., van Ommen, G.-J. B., Baas, F., de Vijlder, J. M., van Duijn, P., and van der Ploeg, M.,** Chromosomal localization of a unique gene by non-autoradiographic *in situ* hybridization, *Nature,* 317, 175, 1985.

108. **Baldino, F., Jr. and Lewis, M. E.,** Non-radioactive *in situ* hybridization histochemistry with digoxigenin-dUTP labeled oligonucleotides, *Methods Neurosci.,* 1, 282, 1989.

109. **Landegren, U., Kaiser, R., Caskey, C. T., and Hood, L.,** DNA diagnostics — molecular techniques and automation, *Science,* 242, 229, 1988.

110. **Urdea, M. S., Warner, B. D., Running, J. A., Stempien, M., Clyne, J., and Horn, T.,** A comparison of non-radioisotopic hybridization assay methods using fluorescent, chemilu-minescent and enzyme labeled synthetic oligodeoxyribonucleotide probes, *Nucl. Acids Res.,* 16, 4937, 1988.

111. **Singer, R. H. and Ward, D. C.,** Actin gene expression visualized in chicken muscle tissue culture by using *in situ* hybridization with a biotinated nucleotide analog, *Proc. Natl. Acad. Sci. U.S.A.,* 79, 7331, 1982.

112. **Singer, R. H., Lawrence, J. B., and Villnave, C.,** Optimization of *in situ* hybridization using isotopic and non-isotopic detection methods, *BioTechniques,* 4, 230, 1986.

113. **Chronwall, B. M., Lewis, M. E., Schwaber, J. S., and O'Donohue, T. L.,** *In situ* hybridization combined with retrograde fluorescent tract tracing, in *Neuroanatomical Tract Tracing Methods,* Heimer, L. and Zaborsky, L., Eds., Plenum Press, New York, 1989, 265.

114. **Fremeau, R. Y., Jr., Lundblad, J. R., Pritchett, D. B., Wilcox, J. N., and Roberts, J. L.,** Regulation of pro-opiomelanocortin gene transcription in individual cell nuclei, *Science,* 234, 1265, 1986.

115. **Milligan, J. F., Groebe, D. R., Witherell, G. W., and Uhlenbeck, O. C.,** Oligoribonucleo-tide synthesis using T7 RNA polymerase and synthetic DNA templates, *Nucl. Acids Res.,* 15, 8783, 1987.

116. **Tecott, L. H., Barchas, J. D., and Eberwine, J. H.,** *In situ* transcription: specific synthesis of complementary DNA in fixed tissue sections, *Science,* 240, 1661, 1988.

117. **Saiki, R. K., Scharf, S., Faloona, F., Mullis, K. B., Horn, G. T., Erlich, H. A., and Arnheim, N.,** Enzymatic amplification of β-globin genomic sequences and restriction site analysis for diagnosis of sickle cell anemia, *Science,* 230, 1350, 1985.

118. **Harbarth, P. and Vosberg, H.-P.,** Enzymatic amplification of myosin heavy-chain mRNA sequences *in vitro, DNA,* 7, 297, 1988.

Chapter 2

DETECTION OF MESSENGER RNAs BY *IN SITU* HYBRIDIZATION WITH NONRADIOACTIVE PROBES

Bertrand Bloch

TABLE OF CONTENTS

I. INTRODUCTION

In situ hybridization has been initiated on the grounds of the nucleic acid hybridization procedures that were developed to detect DNA or RNA target sequences in tissue extracts including solution hybridization and blotting techniques.[1] Since these procedures currently need the use of radioactively labeled DNA or RNA as probes, *in situ* hybridization developed with radioactive probes[2-4] that permit the detection of nucleotide sequences in tissue sections by using macro- or microautoradiography. In many circumstances, the use of radioisotopes for *in situ* hybridization has several disadvantages that limit the potency and the practice of the technique. Efforts, therefore, were made to provide investigators with molecules and procedures that would permit the detection of nucleotide sequences in tissue sections by using nonradioactive probes.[5,6] These probes are directly visible under the microscope or are detected by using a histochemical or immunohistochemical method. The aim here is to present the interest and the limits of the use of nonradioactive probes for messenger RNAs (mRNAs) detection and gene expression analysis by *in situ* hybridization and to describe and discuss the available procedures, the results obtained especially with biotinylated oligonucleotides, as well as the future lines of development in the field.

II. INTEREST AND LIMITS OF NONRADIOACTIVE PROBES FOR *IN SITU* HYBRIDIZATION

The nonradioactive probes have many advantages as compared to radioactive probes: practically, the absence of radioactivity takes away several inconveniences linked to the use of radioisotopes, especially biohazard and the specific organization, precautions, and expenses linked to radioprotection. The use of nonradioactive probes avoids the autoradiographic procedure that is tedious, time consuming, expensive, and difficult to organize on a routine basis. Nonradioactive probes have a long, if not unlimited shelf life and can be used several months, if not years after preparation, something impossible with most radioactive probes (except for the tritiated probes). They can be easily mailed to other laboratories and nonradioactive probes provide results within hours after the completion of the *in situ* hybridization technique, while radioactive probes need days, weeks, and sometimes months of latency.[7-10]

These points are especially critical when contemplating the practice of *in situ* hybridization in a nonspecialized environment such as a pathology laboratory in a hospital, where the analysis must be performed frequently and the results obtained quickly.

When considering the results provided by *in situ* hybridization, the nonradioactive probes also demonstrate their superiority over the radioactive probes: they give a high cellular and subcellular resolution, with a reaction limited to the cell compartment containing the nucleotide target. Indeed, the isotopes with high energy such as ^{32}P or ^{35}S that can provide results in a relatively short time (days or weeks) have a relatively poor resolution. The cells expressing a given mRNA can be specifically identified with these isotopes, but an accurate analysis of the subcellular distribution of the mRNA in them is limited by the diffuse localization of the silver grains signaling the reaction in the emulsion layer.[4,7,8,10] The isotopes with lower energy (^{3}H or ^{125}I) can give a decent resolution but generally require weeks if not months of exposure especially for the detection of the messenger RNAs expressed at a low level.[9-11] In all instances, the subcellular resolution is inferior to that obtained with the nonradioactive probes. The accurate analysis of the subcellular compartmentation of nucleotide sequences such as mRNAs or their immediate precursors demands the use of nonradioactive probes particularly for ultrastructural investigations.

At the moment, it can be considered that the nonradioactive probes present two major disadvantages: first, they are generally less sensitive than the corresponding radioactive probes.[5] This has until now limited the use of nonradioactive probes to the detection of nucleotide sequences abundantly represented in cells: viral genomes,[12] repeated sequences in eukaryote genome,[13] and mRNAs present in high abundance such as those coding for cytoskeletal proteins.[14,15] Recently, improvements in sensitivity have also allowed the detection of hormone and neuropeptide mRNAs with biotinylated oligonucleotides (see below). Presently, nonradioactive probes have found their largest development for virus detection,[12] for which sensitivity is generally not a critical limitation. The second disadvantage of the nonradioactive probes is that they do not permit, at the moment, quantitative *in situ* hybridization because the standardization of the procedure and of the intensity of the reaction is difficult to obtain.[6] The development of quantitation is especially important for analysis of variations of gene activity under experimental, physiological, or pathological conditions by measuring variations in the number of mRNA copies in cells. One must use radioactive probes that permit quantitation for such investigations.[7,16-18]

III. PROCEDURES FOR THE DETECTION OF NUCLEOTIDE SEQUENCES WITH NONRADIOACTIVE PROBES

The literature provides many examples of *in situ* hybridization with nonradioactive probes in tissues and biological specimens (tissue sections, smears, cell cultures, etc.). Nevertheless, few articles are available when considering the detection of mRNAs present in low abundance. A large number of protocols have been proposed and many efforts have been developed to obtain a universal and versatile procedure similar to the fluorescence or peroxidase methods existing for immunohistochemistry. Among these, the most efficient approaches involve a histochemical or immunohistochemical detection of biotin incorporated into the probe.[5,6]

Presently, nonradioactive detection protocols are based on three different approaches:

1. Immunological detection of hybrids with specific antibodies
2. Direct observation of a molecule bound to the probe
3. Detection of a molecule bound to the probe by immunohistochemistry or histochemistry

A. Immunological Detection of Hybrids

Antibodies against DNA/RNA hybrids have been used for *in situ* hybridization to detect either RNA or DNA sequences, especially in polytenic chromosomes.[19] Recently, monoclonal antibodies specific for double stranded DNA have been obtained and their use for *in situ* hybridization contemplated.[20] A direct immunological detection of the hybrids is appealing because it would provide a universal procedure avoiding the labeling of each probe and would simplify the technique, but at the moment, the results obtained have been restricted to a few targets for which sensitivity was not a critical limitation.

B. Direct Observation of a Reactive Molecule Bound to the Probe

Bauman and co-workers[21] developed this approach by attaching a rhodamine derivate to RNA probes in order to detect DNA target sequences. This was also used by Pachman et al.[22] for the detection of the immunoglobulin heavy chain mRNA. At the moment, this procedure is not used outside these groups. Efforts have also been developed to prepare chemiluminescent probes.[23]

C. Histochemical or Immunohistochemical Detection of a Molecule Present in the Probe

Currently, this is the most versatile and universal strategy for nonradioactive detection of nucleotide sequences. Molecules that can be directly or indirectly visualized under the microscope can be tagged to nucleotide probes. These include alkaline phosphatase,[24] biotin,[5,6,12-15,25,26] photobiotin,[6,26] and dinitrophenol.[27] These molecules are attached by a covalent link to the probe (alkaline phosphatase, biotin) or are fixed to a nucleotide (biotin, dinitrophenol), that will be secondarily incorporated into the probe. They are then detected by histochemistry with the appropriate substrate, or by immunohistochemistry. Also, a modified nucleotide (bromodeoxyuridine)[28] can be incorporated into the probe, or created inside the probe by a chemical reaction: acetoxy-acetylaminofluorene modified guanine,[29,30] mercurated DNA,[31] sulfonated cytidine.[32] The modified nucleotides are then specifically recognized by appropriate antibodies. The development of nonradioactive probes implies that the integration of a reporter molecule in the probe is stable, does not significantly alter the hybridization characteristics of it, and that the molecule bound to the probe is accessible and recognized by using a specific detection system. The pioneer work of Ward and colleagues led in the early 1980s to the preparation of deoxynucleotide derivates in which biotin was included through a long allylamine linker arm.[5,13,14,25] Several biotinylated dUTP, dCTP, and UTP derivatives have since been developed and are now available. A careful evaluation of these molecules has demonstrated that their incorporation in double stranded DNA did not significantly alter hybridization properties. These biotin substituted nucleotides are presently the most available and widely used markers. They can be incorporated in probes according to several procedures, and can be integrated in double stranded DNA,[12-15,26] single stranded M13 DNA,[33] complementary RNA,[34,35] and synthetic oligonucleotides.[36-39] Numerous refinements of histochemistry and immunohistochemistry have been used to detect biotin and other molecules: the biotin can be revealed by using antibiotin antibodies and a secondary antibody linked to fluorescein, peroxidase, alkaline phosphatase, or gold particles. Alternatively, biotin can be detected with avidin or better streptavidin, to which is attached any marker. Amplification systems using multistep procedures, streptavidin biotin complexes, or other refinements also improve the sensitivity of the technique.

In these conditions, biotinylated probes have demonstrated efficiency in the detection of a large variety of targets: viral genomes such as papilloma, herpes, cytomegalovirus, and others,[12,40,41] oncogenes,[6,35] and mRNAs. The latter includes the mRNAs coding for cytoskeletal proteins,[14,15,42] transferrin,[34] several hormones, and neuropeptides.[36,38,39,43-45] The use of alkaline phosphatase as a detection system particularly improves the sensitivity of the procedure as compared to peroxidase or fluorescein. This is probably due to the fact that alkaline phosphatase can be reacted several hours with the appropriate substrate, which is generally a mixture of bromochloro-indolyl phosphate and nitroblue tetrazolium.

IV. DETECTION OF HORMONE AND NEUROPEPTIDE mRNAs WITH BIOTINYLATED SYNTHETIC OLIGONUCLEOTIDE PROBES

Radioactively labeled oligonucleotides have been used for several years to detect mRNAs by *in situ* hybridization especially in the nervous system and the endocrine glands.[7,46] The oligonucleotides have several practical interests: they can be easily synthetized by using automatic devices and some are now commercially available. They can be specifically tailored to recognize a limited region of a nucleotide sequence. This is especially useful to explore mRNAs metabolism including maturation and splicing phenomenons by preparing probes that hybridize with a given exon or intron transcript.[36] This can also be profitable by selecting specific probes that allow the specific identification of a target RNA or DNA belonging to families having large homologies[37] (see below).

The oligonucleotides can be labeled with radioactive nucleotides using several procedures. The most common is the tailing procedure that uses terminal deoxynucleotidyl transferase to add radioactive nucleotides at the 3'-end of the probe.[47] Such probes have largely demonstrated their efficiency especially to detect hormone and neuropeptide mRNAs. Their homogeneity favors reproducibility of results. They can be as sensitive as recombinant probes and their use does not require the specific environment and expertise in molecular genetics needed by the preparation and use of recombinant probes. In addition, the availibility of large quantities of oligonucleotides favors the development of internal controls for specificity in the reaction that allow the detection of unspecific binding of the probes to sections. These controls can be performed easily by adding increasing concentrations of synthetic oligonucleotides identical to or different from the probe in the hybridization medium.[7,36] If the reaction is specific, there is a dose-dependent decrease and a disappearance of the staining only when adding the oligonucleotide identical to the probe.

We and others have recently designed procedures that allow the detection of mRNAs[36,38,39,43-45] and viral DNA[37] with biotinylated oligonucleotides by *in situ* hybridization. It appears that several mRNA species can be detected with such probes with a sensitivity approaching the one obtained with the corresponding radioactive probes. Messenger RNAs can be detected in cryostat cut sections[36,39,43,44] (Figures 1 and 2) and paraffin[38] and plastic[45] (Figures 1e and 1f) embedded tissues. Oligonucleotide probes used were single stranded synthetic oligodeoxyribonucleotides 20 to 45 bases long. In most instances, they were labeled at their 3'-end by addition of biotinylated deoxynucleotides with the tailing procedure.[36-38,44]

The addition of nucleotides can be monitored with traces of radioactive nucleotides[36-38,44] or with a histochemical reaction.[39] Alternatively, the biotin can be added at the 5'-end by various chemical procedures.[36,39,44,45,48] In these conditions, the biotin was detected according to: use of the streptavidin-alkaline phosphatase, or a multistep technique involving avidin-biotin complexes and/or anti-avidin antiserum detected with alkaline phosphatase or peroxidase. Papillomavirus DNA can be detected in a 2-h hybridization step with 30 base oligomers permitting differentiation between virus subtypes having large sequence homologies.[37]

Vasopressin and proopiocortin mRNAs have been detected in the rat brain and pituitary by Larsson et al. and Arai et al.[38,39] Our group has personal experience with the detection of several mRNAs in rat and human tissues with this approach.[36,43,44,45] This includes the detection of the neurons expressing several neuropeptide genes in the rat brain, and of the endocrine cells expressing the proopiocortin gene in the pituitary (Figures 1 and 2) and calcitonin gene in medullary thyroid carcinoma.

In our experience, the biotinylated oligonucleotides detected with streptavidin-alkaline phosphatase provided results with a sensitivity similar to the one observed with the corresponding radioactive probes. The reaction product was restricted to the cytoplasm and to process expansions for neurons. The controls performed by addition of unlabeled oligonucleotides demonstrated the specificity of the reaction in human and rat tissue sections.[36]

We have especially studied the vasopressin (AVP) gene expression. In the normal adult rat, the AVP mRNA was detected with several AVP probes in the magnocellular neurons of the supraoptic and paraventricular nucleus which are

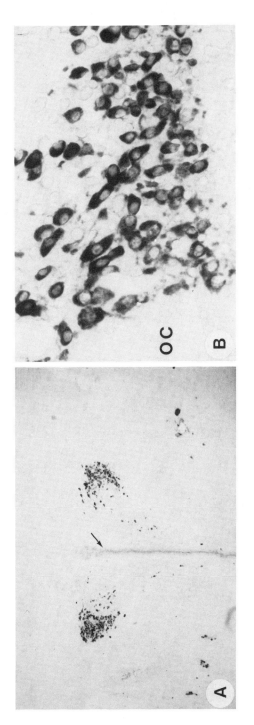

FIGURE 1. Detection of the vasopressin mRNA by *in situ* hybridization with a synthetic oligonucleotide probe containing 10 biotins. Revelation with streptavidin-alkaline phosphatase. (A) and (B) Detection of the vasopressin mRNA in the adult rat brain; paraventricular (A) and supraoptic (B) nucleus of the hypothalamus demonstrate a reaction restricted to the cytoplasm of magnocellular neurons (cryostat cut sections). Arrow = third ventricle. o.c. = optic chiasma. (C) and (D) Detection of the vasopressin mRNA during development at fetal day 19 and 21. (C) shows staining in the supraoptic and paraventricular nuclei. o.c. = optic chiasma. (D) shows staining in neurons that present an immature aspect as compared to (B) (cryostat cut sections). (E) and (F) Detection of the vasopressin mRNA in neurons of the normal (E) and Brattleboro (F) shows that the vasopressin gene deletion present in the Brattleboro rat leads to a peripheral localization of the vasopressin mRNA in the cytoplasm (plastic-embedded semi-thin sections).

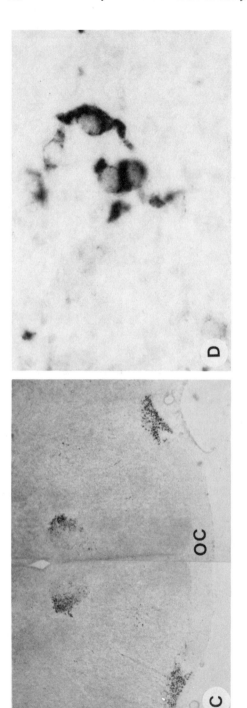

FIGURE 1 continued.

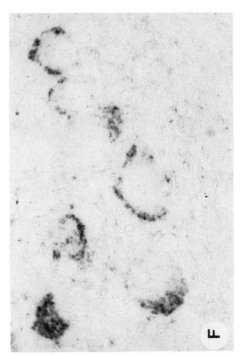

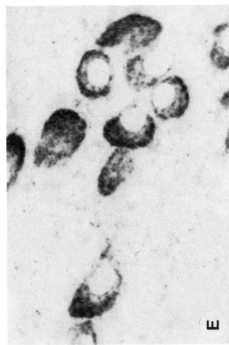

FIGURE 1 continued.

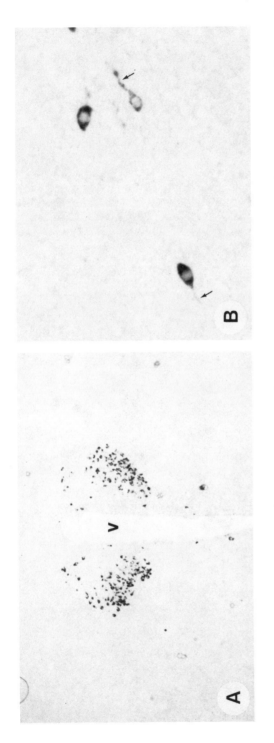

FIGURE 2.   Detection of neuropeptide and hormone mRNAs by *in situ* hybridization with a synthetic oligonucleotide probe. Probes labeled by tailing with biotin dUTP as in References 36 and 44. Adult rat brain and pituitary. Cryostat cut sections. (A) and (B) Detection of oxytocin mRNA in the supraoptic nucleus (A) and LH-RH mRNA in the anterior hypothalamic area (B). V = third ventricle. The probe used to detect the LH-RH mRNA is complementary in the region coding for the gonadotrophin releasing factor-associated peptide (GAP). The arrows point to process containing LH-RH mRNA. (C) Detail of neurons containing somatostatin mRNA in the amygdala. (D) Detection of the proopiocortin mRNA in the cells of the intermediate lobe (IL) and anterior lobe (AL) of the pituitary.

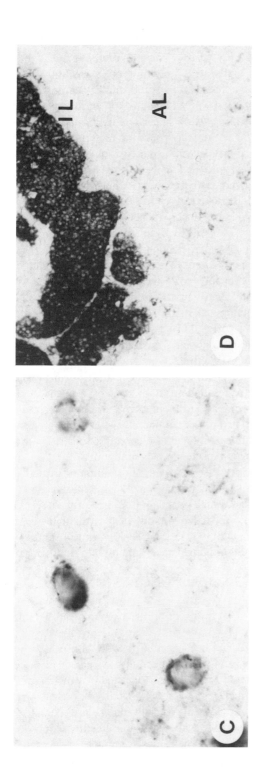

FIGURE 2 continued.

known to contain high levels of AVP mRNA. The use of an oligonucleotide probe containing 10 biotins incorporated at the 5'-end improved sensitivity of the procedure and allowed the reliable detection of parvocellular neurons in the suprachiasmatic nucleus and other hypothalamic areas known to express the AVP gene at a low level.[44,45] This probe was also used to demonstrate the first stages of expression of the AVP gene from day 16 during fetal life in the rat.[44] The neuroblastic appearance of the first neurons and their evolution throughout pre- and postnatal ontogeny clearly appeared with the biotinylated probes while it was not easily detected with the radioactive probes (Figure 1d). The use of a pre-embedding procedure using Vibratome cut sections also allowed the detection of the AVP mRNA in normal and Brattleboro rat neurons with radioactive[49] and biotinylated oligonucleotide probes[46] in semi-thin sections. The combination of plastic embedding and histochemical detection of biotin largely improved cellular and subcellular resolution of the procedure. Simultaneous detection of antigens and mRNAs on this material allowed an accurate cell-by-cell analysis of transcriptional and translational events. Our results demonstrated that in the normal rat, most neurons containing AVP mRNA also contain AVP immunoreactivity; there is no constant correlation between the antigen content and the mRNA level, and some neurons containing AVP mRNA do not contain detectable AVP immunoreactivity. AVP mRNA is present in the cell soma, and also in process expansions and possibly in other cell compartments that could correspond to dendritic spines. In the Brattleboro rats that express a deleted AVP gene, the AVP mRNA has a distribution different from that observed in the normal rat: it is specifically restricted to the peripheral part of the cytoplasm (Figures 1e and 1f). This suggests that AVP gene deletion is associated with a modification in the compartmentation of the corresponding mRNA.[45] Experiments performed in adult normal rats also allowed the detection of oxytocin, somatostatin, and LH-RH neurons (the latter with two probes that recognize either the LH-RH or the LH-RH associated peptide [GAP] part of the mRNA) (Figure 2). Use of probes specific for distinct exons of the calcitonin gene demonstrated that calcitonin and calcitonin-gene related peptide mRNAs are co-transcribed in the same tumoral cells of human medullary thyroid carcinoma.[36]

V. PERSPECTIVES AND CONCLUSION

Nonradioactive probes for *in situ* hybridization are now spreading in all fields of research and diagnostics, especially through the use of biotinylated probes. These probes become more available and easier to use for nonspecialized investigators. Improvements in sensitivity are forthcoming, through an increase in the number of reporter molecules attached to the probe, and the amelioration of the detection systems that could include the use of *in situ* transcription in tissue sections.[50] Oligonucleotides can provide an exquisite molecular recognition inside the target sequences and can be made available in large quantities for many laboratories as a homogenous compound. These make them the "monoclonal antibodies" of *in situ* hybridization that will make it especially profitable to explore mRNA maturation and metabolism, including alternate splicing.

In situ hybridization with nonradioactive probes is compatible with other histological approaches including immunohistochemistry, neuroanatomical tracing, or receptor detection. Transcriptional and translational events can now be investigated together on similar specimens.[45] In the same way, the comparative study of the localization of two mRNAs in the same material can be obtained,[35,51] a goal especially important in neuroanatomy and endocrinology toward understanding the phenomena leading to the co-expression of several hormone or neuropeptide genes in the same cells. Nonradioactive probes are also the appropriate and tools of choice for an accurate analysis of the subcellular compartmentation and metabolism of nucleotide sequences including mRNAs through electron microscopy. Work to establish the detection of Po mRNA in glial cells,[52] mitochondrial RNA in the ovary[53] and gene mapping in chromosomes[54] has borne fruit. The rapid progress observed in the biochemistry of the nonradioactive probes suggests that they will soon compete with, if not replace, the radioactive probes.

ACKNOWLEDGMENTS

The documents presented in this article are the results of several investigations performed with my co-workers, A. F. Guitteny, S. Chouham, C. Mougin, and E. Normand, by using the procedure developed by A. F. Guitteny in the laboratory. I thank R. Teoule, B. Fouqué, and A. Roget from the C. E. N. of Grenoble who prepared some of the biotinylated oligonucleotides used for these investigations. This work was supported by funds from Fondation pour la Recherche Médicale, Ministère de la Recherche et de l'Enseignement Supérieur, and C. I. S. Bioindustrie.

REFERENCES

1. **Maniatis, T., Fritsch, E. F., and Sambrook, J.,** *Molecular Cloning, A Laboratory Manual,* Cold Spring Harbor Laboratory, Cold Spring Harbor, NY, 1982.
2. **John, H.A., Birnstiel, M. L., and Jones, K. W.,** RNA-DNA hybrids at the cytological level, *Nature,* 223, 582, 1969.
3. **Brahic, M. and Haase, A.,** Detection of viral sequences of low reiteration frequency by *in situ* hybridization, *Proc. Natl. Acad. Sci. U.S.A.,* 75, 6125, 1978.
4. **Coghlan, J. P., Penschow, J. D., Hudson, P. H., and Niall, H. D.,** Hybridization histochemistry: use of recombinant DNA for tissue localization of specific mRNA populations, *Clin. Exp. Hypertension,* 6, 63, 1984.
5. **Singer, R. H., Lawrence, J. B., and Villnave, C.,** Optimization of *in situ* hybridization using isotopic and non-isotopic detection methods, *Biofeature,* 4, 230, 1986.

6. **Bresser, J. and Evinger-Hodges, M. J.,** Comparison and optimization of *in situ* hybridization procedures yielding rapid, sensitive mRNA detections, *Gene Anal. Technol.,* 4, 89, 1987.

7. **Uhl, G. R., Zingg, H. H., and Habener, J. F.,** Vasopressin mRNA *in situ* hybridization: localization and regulation studied with oligonucleotide cDNA probes in normal and Brattleboro rat hypothalamus, *Proc. Natl. Acad. Sci. U.S.A.,* 82, 5555, 1985.

8. **Bloch, B., Popovici, T., LeGuellec, D., Normand, E., Chouham, S., Guitteny, A. F., and Böhlen, P.,** *In situ* hybridization histochemistry for the analysis of gene expression in the endocrine and central nervous system tissue: a three-year experience, *J. Neurosci. Res.,* 16, 983, 1986.

9. **Gee, C. E., Chen, C. L. C., Roberts, J. L., Thompson, R., and Watson, S. J.,** Identification of proopiomelanocortin neurones in rat hypothalamus by *in situ* cDNA-mRNA hybridization, *Nature,* 306, 374, 1983.

10. **Hoefler, H., Childers, H., Montminy, M. R., Lechan, R. M., Goodman, R. H., and Wolfe, H. J.,** *In situ* hybridization methods for the detection of somatostatin mRNA in tissue sections using antisense RNA probes, *Histochem. J.,* 18, 597, 1986.

11. **Lewis, M.E., Arentzen, R., and Baldino, J.R.,** Rapid, high-resolution *in situ* hybridization histochemistry with radioiodinated synthetic oligonucleotides, *J. Neurosci. Res.,* 16, 117, 1986.

12. **Brigati, D. J., Myerson, D., Leary, J. J., Spalholz, B., Travis, S. Z., Fong, C. K. Y., Hsiung, G. D., and Ward, D. C.,** Detection of viral genomes in cultured cells and paraffin-embedded tissue sections using biotin-labeled hybridization probes, *Virology,* 126, 32, 1983.

13. **Langer-Safer, P. R., Levine, M., and Ward, D. C.,** Immunological method for mapping genes on Drosophila polytene chromosomes, *Proc. Natl. Acad. Sci. U.S.A.,* 4381, 1982.

14. **Singer, R. H. and Ward, D. C.,** Actin gene expression visualized in chicken muscle tissue culture by using *in situ* hybridization with a biotinated nucleotide analog, *Proc. Natl. Acad. Sci. U.S.A.,* 79, 7331, 1982.

15. **Eng, L. F., Stöcklin, E., Lee, Y. L., Shiurba, R. A., Coria, F., Halks-Miller, M., Mozsgai, C., Fukayama, G., and Gibbs, M.,** Astrocyte culture on nitrocellulose membranes and plastic: detection of cytoskeletal proteins and mRNAs by immunocytochemistry and *in situ* hybridization, *J. Neurosci. Res.,* 16, 239, 1986.

16. **Young, S., Bonner, T., and Brann, M.,** Mesencephalic dopamine neurons regulate the expression of neuropeptide mRNAs in the rat forebrain, *Proc. Natl. Acad. Sci. U.S.A.,* 83, 9827, 1986.

17. **Normand, E., Popovici, T., Onteniente, B., Fellmann, D., Piater-Tonneau, D., Auffray, C., and Bloch, B.,** Dopaminergic neurons of the substantia nigra modulate preproenkephalin A gene expression in the rat striatal neurons, *Brain Res.,* 439, 39, 1988.

18. **Rogers, W., Schwaber, J., and Lewis, M.,** Quantitation of cellular resolution *in situ* hybridization histochemistry in brain by image analysis, *Neurosci. Lett.,* 82, 315, 1987.

19. **Rudkin, G. T. and Stollar, B. D.,** High resolution detection of DNA-RNA hybrids *in situ* by indirect immunofluorescence, *Nature,* 265, 472, 1977.

20. **Huang, C. M., Huang, H. J. S., Glembourtt, M., Liu, C. P., and Cohen, S. N.,** Monoclonal antibody specific for double-stranded DNA: a non-radioactive probe method for detection of DNA hybridization, in *Rapid Detection and Identification of Infectious Agents,* Academic Press, New York, 1985, 257.

21. **Bauman, J. G. J., Wiegant, J., and Van Duijn, P.,** Cytochemical hybridization with fluorochrome-labeled RNA. II. Applications, *J. Histochem. Cytochem.,* 29, 238, 1981.

22. **Pachman, V., Pech, M., Pachman, U., and Dormer, P.,** Identification of messenger RNA coding for the constant fragment of the heavy chain with cloned DNA in single cells by *in situ* hybridization, *Blut,* 46, 107, 1983.

23. **Heller, M. J. and Morrison, L. E.,** Chemiluminescent and fluorescent probes for DNA hybridization systems, in *Rapid Detection and Identification of Infectious Agents,* Academic Press, New York, 1985, 245.

24. **Jablonski, E., Moomaw, E. W., Tullis, R. H., and Ruth, J. L.,** Preparation of oligodeoxynucleotide-alkaline phosphatase conjugates and their use as hybridization probes, *Nucl. Acids Res.,* 14, 6115, 1986.

25. **Langer, P. R., Waldrop, A. A., and Ward, D. C.,** Enzymatic synthesis of biotin-labeled polynucleotides: novel nucleic acid affinity probes, *Proc. Natl. Acad. Sci. U.S.A.,* 78, 6633, 1981.

26. **Forster, A., McInnes, J. L., Skingle, D. C., and Symons, R.,** Non-radioactive hybridization probes prepared by the chemical labelling of DNA and RNA with a novel reagent, photobiotin, *Nucl. Acids Res.,* 13, 745, 1985.

27. **Shroyer, K. and Nakane, P.,** Use of DNP-labeled cDNA for *in situ* hybridization, *J. Cell Biol.,* 97, 377a, 1983.

28. **Niedobitek, G., Finn, T., Herbst, H., Bornhöft, G., Gerdes, J., and Stein, H.,** Detection of viral DNA by *in situ* hybridization using bromodeoxyuridine-labeled DNA probes, *Am. J. Pathol.,* 131, 1, 1988.

29. **Tchen, P., Fuchs, R. P. P., Sage, E., and Leng, M.,** Chemically modified nucleic acids as immuno-detectable probes in hybridization experiments, *Proc. Natl. Acad. Sci. U.S.A.,* 81, 3466, 1984.

30. **Landegent, J. E., Jansen In De Wal, N., Baan, R. A., Hoeijmakers, J. H. J., and Van Der Ploeg, M.,** 2-Acetylaminofluorene-modified probes for the indirect hybridocytochemical detection of specific nucleic acid sequences, *Exp. Cell. Res.,* 153, 61, 1984.

31. **Hopman, A. H. N., Wiegant, J., and Van Duijn, P.,** A new hybridocytochemical method based on mercurated nucleic acid probes and sulfhydryl-haptem ligands. II. Effects of variations in ligand structure on the *in situ* detection of mercurated probes, *Histochemistry,* 84, 179, 1986.

32. **Duthil, B., Bebear, C., Taylor-Robinson, D., and Grimont, P.,** Detection of *Chlamydia trachomatis* by *in situ* hybridization with sulfonated total DNA, *Ann. Inst. Pasteur,* 139, 115, 1988.

33. **Varndell, I. M., Polak, J. M., Sikri, K. L., Minth, C. D., Bloom, S. R., and Dixon, J. E.,** Visualisation of messenger RNA directing peptide synthesis by *in situ* hybridization using a novel single-stranded cDNA probe. Potential for the investigation of gene expression and endocrine cell activity, *Histochemistry,* 81, 597, 1984.

34. **Morales, C., Hugly, S., and Griswold, M. D.,** Stage dependent levels of specific mRNA transcripts in Sertoli cells, *Biol. Reprod.,* 36, 1035, 1987.

35. **Evinger-Hodges, M. J., Blick, M., Bresser, J., and Dicke, K. A.,** Comparison of oncogene expression in human normal bone marrow and leukemia, *Ann. N.Y. Acad. Sci.,* 511, 284, 1987.

36. **Guitteny, A. F., Fouqué, B., Mougin, C., Téoule, R., and Bloch, B.,** Histological detection of messenger RNAs with biotinylated synthetic oligonucleotide probes, *J. Histochem. Cytochem.,* 36, 563, 1988.

37. **Cubie, H. A. and Norval, M.,** Synthetic oligonucleotide probes for the detection of human papilloma viruses by *in situ* hybridization, *J. Virol. Methods,* 20, 239, 1988.

38. **Larsson, L., II, Christensen, T., and Dalboge, H.,** Detection of proopiomelanocortin mRNA by *in situ* hybridization, using a biotinylated oligodeoxynucleotide probe and avidin-alkaline phosphatase histochemistry, *Histochemistry,* 89, 109, 1988.

39. **Arai, H., Emson, P. C., Agarwal, S., Christodoulou, C., and Gait, M. J.,** *In situ* hybridization histochemistry: localisation of vasopressin mRNA in rat brain using a biotinylated oligonucleotide probe, *Mol. Brain Res.,* 4, 63, 1988.

40. **Burns, J., Redfern, D. R. M., Esiri, M. M., and McGee, J. O. D.,** Human and viral gene detection in routine paraffin embedded tissue by *in situ* hybridisation with biotinylated probes: viral localisation in herpes encephalitis, *J. Clin. Pathol.,* 39, 1066, 1986.

41. **Beckmann, A. M., Myerson, D., Daling, J. R., Kiviat, N. B., Fenoglio, C. M., and McDougall, J. K.,** Detection and localization of human papilloma virus DNA in human genital condylomas by *in situ* hybridization with biotinylated probes, *J. Med. Virol.,* 16, 265, 1985.

42. **Lawrence, J. B. and Singer, R. H.,** Intracellular localization of messenger RNAs for cytoskeletal proteins, *Cell,* 45, 407, 1986.

43. **Bloch, B., Guitteny, A. F., Normand, E., Chouham, S., Le Moine, C., Fouque, B., and Teoule, R.,** Histological detection of neuropeptide messenger RNAs with radioactive and biotinylated synthetic oligonucleotides, in *Wenner Gren International Symposium Series,* Vol. 3, Macmillan, London, 1989, 23.

44. **Bloch, B., Guitteny, A. F., Chouham, S., Mougin, C., Fouque, B., Roget, A., and Teoule, R.,** Topography and ontogeny of the neurons expressing vasopressin, oxytocin, and soma- tostatin genes in the rat brain: an analysis using radioactive and biotinylated oligonucleotides, *Cell. Mol. Neurobiol.,* in press.

45. **Guitteny, A. F., Fouque, B., Teoule, R., and Bloch, B.,** Vasopressin gene expression in normal and Brattleboro rat: histological analysis in semi-thin sections with biotinylated oligonucleotide probes, *J. Histochem. Cytochem.,* 37, 1479, 1989.

46. **Lewis, M. E., Sherman, T. G., and Watson, S. J.,** *In situ* hybridization histochemistry with synthetic oligonucleotides. Strategies and methods, *Peptides,* 6, 75, 1985.

47. **Roychoudhury, R. and Wu, R.,** Terminal transferase-catalyzed addition of nucleotides to 3′ termini of DNA, in *Methods in Enzymology,* Academic Press, London, 1980, 43.

48. **Chollet, A. and Kawashima, E. H.,** Biotin-labeled synthetic oligodeoxyribonucleotides: chemical synthesis and uses as hybridization probes, *Nucl. Acids Res.,* 13, 1529, 1985.

49. **Guitteny, A. F., Böhlen, P., and Bloch, B.,** Analysis of vasopressin gene expression by *in situ* hybridization and immunohistochemistry on semi-thin sections, *J. Histochem. Cyto- chem.,* 36, 1373, 1988.

50. **Tecott, L. H., Barchas, J., and Eberwine, J.,** *In situ* transcription: specific synthesis of complementary DNA in fixed tissue sections, *Science,* 240, 1661, 1988.

51. **LeMoine, C., Normand, E., Guitteny, A. F., Foque, B., Teoule, R., and Bloch, B.,** Dopamine receptor gene expression by enkephalin neurons in the rat forebrain, *Proc. Natl. Acad. Sci. U.S.A.,* 230, 87, 1990.

52. **Webster, H. de F., Lamperth, L., Favilla, J. T., Lemke, G., Tesin, D., and Manuelidis, L.,** Use of a biotinylated probe and *in situ* hybridization for light and electron microscopic localization of P_0 mRNA in myelin-forming Schwann cells, *Histochemistry,* 86, 441, 1987.

53. **Binder, M., Tourmente, S., Roth, J., Renaud, M., and Gehring, W. J.,** *In situ* hybridiza- tion at the electron microscope level: localization of transcripts on ultrathin sections of Lowicryl K4M-embedded tissue using biotinylated probes and protein A-gold complexes, *J. Cell. Biol.,* 102, 1646, 1986.

54. **Hutchison, N. J., Langer-Safer, P. R., Ward, D. C., and Hamkalo, B. A.,** *In situ* hybridization at the electron microscope level: hybrid detection by autoradiography and colloidal gold, *J. Cell. Biol.,* 95, 609, 1982.

Chapter 3

IN SITU HYBRIDIZATION IN CELLS AND TISSUE SECTIONS: A STUDY OF MYELIN GENE EXPRESSION DURING CNS MYELINATION AND REMYELINATION

Craig A. Jordan

TABLE OF CONTENTS

I. INTRODUCTION

Over the past 15 years, molecular genetic studies have become common and often essential in many diverse research disciplines. The development of recombinant DNA technology initiated the ability to do these types of studies and has produced sensitive methods to analyze gene structure, expression, regulation, and function. One such technique, *in situ* hybridization, allows gene expression to be studied at the level of a single cell. This powerful technique has been utilized to assign chromosomal locations of DNA gene sequences,[1] to localize RNA transcripts within individual cells at the light and electron microscope level,[2-4] to detect activity of promoter or enhancer sequences used in transfected cell or transgenic mouse expression systems,[5,6] and to identify cell types which express developmentally or differentiationally regulated genes.[7,8]

In situ hybridization has been performed on chromosomal preparations, tissue sections, and cultured cells. This chapter discusses *in situ* hybridization on cultured cells and tissue sections which require different preparation for *in situ* hybridization and can be utilized to address either different or complementary questions. *In situ* hybridization on tissue sections allows the examination of natural *in vivo* levels of expression for a specific gene transcript and localization of these transcripts to specific anatomical region(s) where they are preferentially expressed. When cells are dissociated and grown in culture, the anatomical landmarks are lost and the level of gene expression is often reduced. However, these cells can be purified and manipulated by the addition of growth factors, mitogens, or co-culture with other cell types in order to evaluate the effect of such interactions on gene expression. Duplicate cell cultures can also be characterized on the basis of specific histologic stains, cellular morphologies, or expression of specific cell markers in parallel to *in situ* hybridization.

II. TECHNICAL CONSIDERATIONS

A. Specimen Preparation

A variety of cell preparations can be analyzed by *in situ* hybridization for the expression of specific transcripts. Cells in suspension, either from cultures or dispersed from tissues, can be cytocentrifuged onto slides, [9,10] embedded in a freezable mounting medium such as OCT and sectioned,[11] or deposited as cell smears or tissue touch preps. Adherent cells can be grown on coverslips which are then glued to slides with a strong glue such as Autoslide adhesive for easier handling.[7,12,13] More recently, adherent cells were simply seeded into four well chambered slides (Lab-tek). The chambers are helpful during fixation and pretreatment of the monolayers, but are easily removed prior to hybridization. This also allows multiple probes, cell types, and/or cell treatments to be used on the same slide. The slides or coverslips used for these procedures usually are first coated with polylysine,[9,12-14] gelatin,[8,15] laminin,[16] or Denhardt's solution.[7,9,17] Cultured cells grown on plastic can be used for *in situ* hybridization, but they require more tedious handling, are not compatible with chloroform or xylene treatments, and the plastic has poor optical qualities when compared to glass. Cell preparations are often used immediately for *in situ* hybridization, but may be stored for extended periods before analysis. The preparations are usually fixed and dehydrated, sometimes stored dry at 4°C[26] or more commonly in 70% ethanol at 4°C.[10,13,15]

Tissue for *in situ* hybridization can be from perfused animals, immersion fixed biopsy or autopsy specimens, or dissected fresh and quickly frozen. Some investigators prefer fresh frozen tissues over perfused tissues and cite better signal and lower background probe binding.[18,19] The perfused tissues perform well for *in situ* hybridization, can be readily immunostained,[20,21] and are optimal for studies requiring good morphologic preservation.[22,23] Perfused tissue can be cryoprotected, snap frozen, and sectioned, or embedded prior to sectioning. A variety of embedding compounds have been used successfully: paraffin, gelatin, Lowicryl, and 2.3 *M* sucrose.[16,21-23] In our laboratory, tissues from animals perfused with 4% paraformaldehyde are immersed overnight in 15% sucrose for cryoprotection, and then snap frozen in liquid nitrogen (see Protocols). When such tissues are sealed in airtight containers and stored at −70°C, they retain their mRNA for greater than 1 year. Cryostat sections cut from these blocks of tissue retain their mRNA for at least several months when stored in a nondefrosting freezer (−20°C). Long-term, stable storage allows tissues collected from different experiments or during a long time course to be directly compared to each other during the same *in situ* hybridization analysis.

Poor retention of tissue sections on slides can sometimes be a problem because of harsh treatment, extensive washing, and relatively high temperatures used following hybridization. A variety of methods are used to ensure adherence of tissue sections to glass slides. After sulfuric acid cleaning, rinsing, and baking to destroy RNase activity, slides are coated with one of the following: Histostik,[15] poly-L- or poly-D-lysine,[13,14] 3-aminopropyl-triethoxysilane activated with an aldehyde fixative,[16,24] Denhardt's solution,[7,17] gelatin,[8,11,19] or a combination.[18]

B. Pretreatment of Cells and Tissues

Pretreatment of cells and tissues serves to increase the accessibility of the target nucleic acid to the probe while minimizing nonspecific adherence of the probe, and maintaining as much morphological detail as possible. The type of tissue, the degree of cellular protein cross-linking following fixation, and the size of the probe will all influence which pretreatments are required. For these reasons it is necessary for investigators to test a number of variables in order to optimize *in situ* detection of mRNA in their system.[9,15]

We have found that cell culture preparations require more rigorous pretreatments than tissue sections (see Protocols section). Fixed cell culture preparations are treated with 0.05% Triton X-100 to disrupt membranes followed by 0.2 N HCl in order to deproteinize the cells and to reduce nonspecific binding of nucleotides.[16] The tissue culture cells are further deproteinized enzymatically with predigested proteinase K (predigestion destroys RNase activity) which is followed by rinses containing 2 mg/ml glycine to stop the reaction. Enzymatic digestion must be tested to determine the optimal concentration of enzyme for a particular cell preparation. Proteinase K is usually effective within the range of 1 to 10 µg/ml.[15,25] Other proteolytic enzymes such as pronase [18,26] can also be used. At this point, cell cultures are fixed again in 4% paraformaldehyde.

We find that tissue sections do not require the permeabilizing and deproteinizing steps just mentioned, most likely because many cells are sliced open during the sectioning process. The types of tissue we use (central nervous system) and the small size of our probes (48 to 300 nucleotides) may also play a role. Some laboratories do use similar pretreatments on tissue sections,[9,16,17,27] while other laboratories find they are not necessary[13,28] or that such treatments actually reduce RNA retention within the cells.[15,28] Again, such differences can often be understood if one considers the type of fixative used and the relative degree of cross-linking which results, the size of the probe, or the type of tissue. Direct comparison of a number of different fixatives and how they affect RNA accessibility and probe binding has been addressed.[9,15,28]

The final steps of pretreatment apply to tissue culture preparations and to tissue sections. Cryostat sections are brought to room temperature and fixed in 4% paraformaldehyde.[8,19] Paraffin sections must be deparaffinized and rehydrated.[20,25] The slides are treated with 0.25% acetic anhydride which neutralizes positive charges on the specimens and slides, thus reducing electrostatic binding of probes which would contribute to background. Slides are then dehydrated in graded ethanols and air dried. During the dehydration phase, we routinely delipidate spinal cord or brain sections in chloroform to avoid nonspecific probe binding to the large amount of lipid present in these tissues. This treatment is not necessary for tissue culture specimens and must be avoided when cells are on plastic dishes.

An additional type of *in situ* hybridization pretreatment designed to reduce nonspecific probe binding is prehybridization. This consists of treating the tissues or cells with all of the ingredients of the hybridization mixture except the probe. The mixture routinely contains bovine serum albumin, salmon sperm DNA, and yeast tRNA in appropriate salt and buffer conditions. During prehybridization, any nonspecific binding that occurs on the cells or tissue will occur before the probe is introduced at the hybridization step. In many laboratories, this step is now omitted as unnecessary,[8,13,19] while others continue to find it beneficial.[17,18,23] Recently, the addition of nonlabeled thio α-UTP to the prehybridization mixture was shown to reduce background occurring with ^{35}S labeled probes.[16,29] Pretreatment to reduce background may prove even more important in attempts to detect lower abundance messages which require longer exposures.

C. Probes

The majority of probes used for *in situ* hybridization are labeled with a radioactive isotope and detected by autoradiography. Over the past several years, nonisotopically labeled probes containing biotin molecules have been developed.[15,30] These probes are detected by histochemical techniques which are faster and safer, but have less sensitivity. However, the appeal of these probes is evidenced by the increased number of laboratories using them and attempts to develop other nonisotopic detection systems such as digoxigenin.[31]

There are a variety of characteristics which should be considered in selecting or designing a probe for *in situ* studies. Penetration of the probe into the tissue in order to hybridize with the appropriate target RNA is essential. Penetration of the probe into the tissue will depend on the type of tissue, the type of fixative used, and the degree of permeabilization obtained by pretreatments, but in general, penetration occurs more readily with smaller probes. Once the probe has penetrated the tissue it should recognize the target RNA with maximal specificity. High specificity requires a high degree of sequence homology between the probe and the target RNA, but not with other uninteresting RNA species. Careful selection of the probe sequence is critical in situations where only one of several closely related RNA transcript forms is to be detected, or where a probe from one species of animal is used to detect a form from a different species with reduced sequence homology. Maximal specificity of the probe for the target sequence will also increase the theoretical hybridization strength and allow posthybridization washes to be performed under more stringent conditions. Increased length and increased G/C content of the probe are other variables which can increase the theoretical hybridization strength, but potential tradeoffs must be considered. Increasing the probe length may decrease penetration into the tissue and may require using regions of sequence with reduced homology or which will hybridize to RNA sequences other than the intended target sequences. Increased G/C content results in stronger hybrid formation but may result in increased background when too far above 50%. The type of nucleic acid used for the probe can also influence the strength of the resulting hybrid. RNA:RNA hybrids formed from the use of cRNA probes display a higher melting temperature than equivalent DNA:RNA hybrids.[25]

The sensitivity of the probe is another important characteristic and it is increased with increasing hybridization affinities and with high specific activities. The specific activity of a particular probe is a function of how many radioactive residues are added or incorporated, and the specific energy of the isotope used ($^{32}P \gg ^{35}S \gg ^{125}I > ^{3}H$). Probes with high specific activities are more easily detected and can reduce background binding since they can be used at lower concentrations during hybridization. Unfortunately, probes of very high specific activity are expensive to make and are less stable. The small degraded radiolabeled molecules can bind nonspecifically and contribute to background. In addition, the resolution of detection is inversely related to the energy of emission for each of the isotopes, such that even resolution at the cellular level is poor for ^{32}P, while ^{3}H is the isotope of choice for detailed subcellular resolution. As a compromise, most investigators use ^{35}S to label their *in situ* probes. This allows good resolution at the cellular level while the relatively high activity allows for shorter exposures.

We currently use either single-stranded synthetic oligonucleotide probes 3'-

end-labeled with ^{35}S-dATP or single-stranded cRNA probes which are labeled with ^{35}S-UTP during *in vitro* transcription of a plasmid containing the cloned cDNA fragment downstream from an RNA polymerase binding site (see Protocols). Since both types of probes are single stranded they do not require denaturation and can only anneal with target mRNA sequences, not with contaminating plus strands which are present in double stranded probe preparations. The synthetic oligonucleotide probes can be produced and labeled quite rapidly for any known sequence without cloning and their small size is ideal for penetration into tissues. At the time of synthesis, these probes can easily be designed to specifically detect particular exons or other limited sequences. By designing probes of identical length for different gene transcripts, comparable quantitation of binding can be obtained for each probe. However, low abundance messages may be difficult to detect with probes representing only a small portion of the target coding sequence. These transcripts are best detected using probes corresponding to large regions. This requires cloning. Once a DNA clone is available, it can be used to generate a cRNA probe which is then alkaline hydrolyzed into smaller fragments for optimal tissue penetration. An additional benefit of this type of probe is that much of the background probe binding can be digested with single strand specific RNAses which will not affect the hybrids.

III. *IN SITU* HYBRIDIZATION OF CNS TISSUES

This laboratory has used *in situ* hybridization of myelin forming cells both *in vitro* and *in vivo* in order to analyze the developmental expression of myelin specific genes in the nervous system. In the central nervous system (CNS) myelin is produced by oligodendrocytes. These multipolar glial cells synthesize, transport, and integrate myelin components into the membrane of their processes which then wrap around axons and form multilamellar myelin sheaths.[32,33] Myelin is essential for rapid saltatory conduction of electrical impulses along nerve fibers. Consequently, demyelination can result in neurologic dysfunction, evident in multiple sclerosis.[34]

CNS myelin is composed of four major proteins: proteolipid protein (PLP), myelin basic protein (MBP), 2′,3′-cyclic nucleotide 3′-phosphohydrolase (CNP), and myelin-associated glycoprotein (MAG). PLP is the major protein of CNS myelin (50% of total protein).[35,36] It is an acetylated, hydrophobic protein which is synthesized on rough endoplasmic reticulum and is thought to have an essential structural role in myelin compaction.[37] MBP (30% of total protein) is synthesized on free ribosomes and is involved in compaction of the major dense line of myelin.[38,39] CNP (5% of total protein) is named for its enzymatic activity in other tissues but as yet its function in myelin is unknown.[36] CNP, similar to MAG, is not found in compact myelin.[40,41] MAG (1% of total myelin) shares structural similarity with many membrane molecules of the immunoglobulin superfamily[42] and appears to have a role in initiation of oligodendrocyte-axon interactions.[22,43] These myelin proteins are under tightly regulated and coordinated genetic control such that all four proteins are expressed almost synchronously in oligodendrocytes within CNS tracts at a time shortly preceding myelination.[44-47]

The recent cloning and sequencing of the myelin protein gene sequences has led to the identification of multiple mRNA species which encode the different protein isoforms for each of these four myelin proteins.[48-55] Alternative splicing [48,49,51,54-56] appears to be a general mechanism by which the different mRNA species are generated for each myelin gene. One can use nucleic acid probes specific for each myelin gene to study gene expression during myelination and remyelination. Use of exon specific probes allows analysis of transcripts which encode specific isoforms of a particular protein. Over the years we have studied neonatal myelination and remyelination following viral induced demyelination and have refined our techniques to detect different types of transcripts during these differentiation events.

A. Primary Oligodendrocyte Cultures

Early studies of cultured rat oligodendrocytes from a variety of CNS regions indicated that these dispersed cells still expressed MBP in a developmentally regulated pattern.[57] Subsequent isolation of a cDNA clone for mouse MBP in this laboratory by Zeller et al.[7] made it possible to examine the expression of this gene at the level of transcription, and to directly correlate mRNA with protein expression. This question was addressed in primary cultures of rat brains.[7] These cultures were initiated from fetal (15 to 17 d gestation) or newborn (0 to 4 d postnatal) rat brains. In other experiments, cultures were enriched for oligodendrocytes, seeded onto poly-L-lysine or laminin-treated coverslips, and assayed for MBP-mRNAs. Parallel cultures were examined for MBP or galactocerebroside (GC) markers at times corresponding to different postnatal ages. GC is a glycolipid expressed by maturing oligodendrocytes and appears prior to MBP. Protein or glycolipid expression was determined by immunofluorescence staining for MBP and GC or glial fibrillary acidic protein (GFAP, a specific marker for astrocytes). *In situ* hybridization was performed at 30°C overnight with 3×10^5 cpm/coverslip of the 3′-labeled cDNA probe which had been tailed with ^{35}S-dATP.

Zeller et al. compared MBP mRNA expression to MBP protein levels in primary brain cultures and in oligodendrocyte enriched cultures at equivalent days postnatal. MBP-specific mRNA was detected in scattered cells in cultures corresponding to 6 d postnatal. These cells had an eccentric nucleus, as described for immature oligodendrocytes. The level of MBP-specific mRNA expression remained high through day 15 postnatal. Detection of MBP-specific mRNA preceded fluorescent detection of MBP by 2 d, and at that time, the percentage of MBP-mRNA positive cells was higher than the percentage of cells expressing MBP. During the second week of culture, the percentages of cells positive by immunofluorescence and *in situ* were approximately equal, and at later times, the percentage of protein positive cells was greater than those expressing MBP-mRNA. A reduced ratio of message to protein would be expected as protein levels plateaued and transcripts were only required to maintain protein levels. Frequently, MBP-mRNA positive oligodendrocytes of older cultures were found to have *in situ* hybridization grains associated with the long processes which developed with maturation of oligodendrocytes (Figure 1). The presence of MBP-mRNA in the cellular processes is thought to be related to its translation on free ribosomes.

These studies showed that the developmental regulation of MBP-mRNA occurs *in vitro* and that this regulation transpires in the absence of neurons, the targets of oligodendrocyte myelination. The *in vitro* timetable corresponds very closely to the timing of MBP-mRNA expression in brain[7] and MBP synthesis *in vivo*,[58] supporting the prediction that MBP expression is regulated at the level of transcription.

In a more recent study by Holmes et al.,[14] primary mixed glial cell cultures from 1 to 2 d old rat brains were analyzed for mRNAs of MBP as well as GFAP and glycerol phosphate dehydrogenase (GPDH), another glial cell marker. In this study, parallel mixed brain cell cultures were hybridized *in situ* with a probe for one of these three glial transcripts. Expression of the three different transcripts was correlated with histological, morphological, and positional characteristics of cell types in the parallel cultures. Different cell populations were shown to express either mRNA for the astrocyte marker GFAP or mRNAs for the oligodendrocyte markers GPDH and MBP. The expression of MBP-mRNA increased rapidly after day 8 and was found even in processes of cells by day 16. However, the cells expressing MBP-mRNA only partially overlapped with the GPDH-mRNA expressing cells. Thus, distinct populations existed which expressed only one of these oligodendrocyte-specific messages. This study confirms previous studies on *in vitro* MBP-mRNA expression and compares it with expression of other lineage-specific mRNAs. Additionally, these findings indicate that oligodendrocytes may be divided into subgroups based on MBP and GPDH expression.

B. Normal Myelinating CNS Tissues

The successful *in vitro* demonstration of cultured oligodendrocytes expressing myelin-specific mRNAs spurred the examination of mRNA expression *in vivo*. In our laboratory Kristensson et al. analyzed the expression of MBP-mRNA in sections of developing rat brain.[17] Sprague-Dawley rats (3 to 60 d old) were anesthetized and perfused. The brains were dissected, post-fixed, infiltrated with 15% sucrose, frozen in liquid nitrogen, and sectioned on a cryostat (10 μm). The expression of MBP-mRNA, detected by *in situ* hybridization, was compared to protein expression which was assayed by immunofluorescence staining on parallel sections. In initial experiments, small (50 to 100 nucleotides) ^{35}S-tailed cDNA probes were found to be better than ^{35}S-nick translated probes which were longer (800 to 1000 nucleotides) and generally of lower specific activity.

MBP-mRNA was detectable in various regions of the CNS white matter

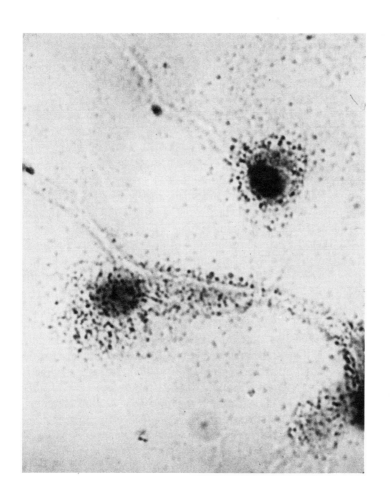

FIGURE 1. *In situ* hybridization of an oligodendrocyte enriched culture with the MBP-cDNA probe. Probe is bound over the cell bodies and along the cellular processes of the differentiated oligodendrocytes, as indicated by the accumulation of silver grains over these regions. (Original magnification × 600. From Zeller, N. K. et al., *J. Neurosci.*, 5, 2955, 1985. With permission.)

during myelination. The MBP-mRNA was detected in large amounts in actively myelinating bundles of young animals, but these levels were reduced in older animals. Levels of MBP-mRNA were measured by counting silver grains in the tectospinal tract of the medulla oblongata and in the corpus callosum of the cerebrum. Message was detected earlier in the medulla oblongata than in the cerebrum, as would be predicted from the caudal to rostral progression of CNS myelination. The medulla oblongata had peak expression in 7-d-old rats while the cerebrum had peak levels in 17-d-old rats, correlating with the rough estimates of peak brain myelination-based northern analysis of messages or immunocytochemical detection of proteins.[58,59]

Subsequently, sequences for the other myelin protein genes became available and we analyzed their expression in comparison to that of MBP.[8] Probes were developed for PLP, MAG, and CNP. Experiments were also designed to determine if subclasses of MBP transcripts could be differentially detected *in vivo*. This was accomplished by construction of synthetic oligonucleotide probes which were complementary to sequences of specific MBP exons. Previous work[49,51,60] has indicated that sequences encoded by MBP exon 2 are present in some mRNA forms but spliced out of other forms. These exon 2-containing forms of MBP predominate early during myelination but are reduced in adults[61,62] where the majority of MBP transcripts lack exon 2 information. Oligonucleotide probes were synthesized for MBP exon 2 and exon 1. The exon 1 probe detects all forms of MBP transcripts, including those containing exon 2, since exon 1 is not spliced out of any MBP-mRNA form.

The spinal cord was chosen for these studies[8] because it presents a uniform and comparable morphology through the many sequential sections required for the variety of nucleic acid probes and histologic stains used. The mouse spinal cord is also the site of demyelination and remyelination in certain experimental models which we wanted to examine for changes in myelin-specific gene expression. To provide an accurate comparison of transcript levels for the different myelin-specific genes, probe binding was quantitated by densitometry of individual experiments in which the only variables were tissue age and probe specificity. In this study we also compared two types of single stranded probes, synthetic oligonucleotide DNA probes and cRNA probes, which were alkaline hydrolyzed to yield 200 to 400 nucleotide fragments. Both types of probes gave good results, but the synthetic oligonucleotide probes had a slightly lower background, were available for all four genes, and allowed more rigorous comparison between the different probes because of their uniform length. The oligonucleotides (48 to 50 nucleotides) were 3′-end labeled with ^{35}S-dATP to a specific activity of 2 to 3×10^8 cpm/µg. Tissues were prepared as described in the section "Protocols".

High levels of all four myelin-specific mRNAs were found in the spinal cord white matter of 8-d-old mice, the earliest age analyzed (Figure 2). Comparison of message levels for the different genes at increasing times postnatal indicated two groups. Total MBP (assayed with the exon 1 probe), PLP, and CNP transcript levels remained high between the ages of 8 to 20 d before starting to decline to adult levels. However, MAG transcripts and the subset of MBP transcripts which contained exon 2, were already declining in abundance by 20 d postnatal. These patterns indicate that although myelin gene transcripts emerge at about the same time in one specific region, expression of the different transcripts is regulated independently at later stages of development. The rapid decline in levels of MAG and exon 2-containing MBP transcripts after the initiation of myelination supports the hypothesis that they are critically required during early stages of differentiation or myelination. MAG has been implicated in initial recognition and adhesion between oligodendrocyte processes and the axons which they ensheath.[22,43] The exon 2 region of MBP codes for a string of amino acids whose folding might be critical for interaction with PLP and/or myelin lipids of apposing membranes during early myelinating events.[62]

Specific transcripts for all four myelin genes were also demonstrated in gray matter regions of the spinal cord but reached peak levels later in development (20 d). The gray matter expression of these oligodendrocyte-specific messages was demonstrated over myelinated fiber tracts and satellite oligodendrocytes associated with motor neurons of the ventral horn. The delayed appearance of these messages indicates that myelination of spinal cord gray matter is delayed relative to white matter. When oligodendrocytes were immunostained for CNP protein and counted in these gray matter regions they were found to increase between 8 and 20 d postnatal. This indicated that oligodendrocytes of the gray matter are generated or mature later than those in the white matter. Studies of oligodendrocyte progenitor cells of the optic nerve indicate that differences in the timing of migration or mitosis of these cells might account for different myelinating schedules of specific CNS tracts.[63]

Identification of the cell type in which transcripts were expressed was easier because of the low background binding of these ^{35}S-oligonucleotide probes and was most clearly seen in scattered satellite oligodendrocytes in the gray matter. PLP and MAG probes were localized over perinuclear regions of the cells and not over processes. This perinuclear distribution of PLP-mRNA has been demonstrated in other laboratories.[27,64] In one of these studies,[27] mRNA for P_o, the major myelin protein in the PNS, was also found to have a tight perinuclear localization in myelinating Schwann cells (also see References 23 and 65). This perinuclear distribution would be expected for mRNAs which are translated on rough endoplasmic reticulum as is thought to be the case for PLP and P_o. MAG probe binding was also exclusively associated with oligodendrocyte perikarya. The pattern of MBP probe binding was found to be quite different from the other myelin-specific probes. In well-myelinated regions, the diffuse probe binding could not be associated with a specific cell body. This could be explained by transport of MBP-mRNA into cell processes, where they have been demonstrated *in vitro*.[7] CNP is synthesized on free ribosomes[54,66] similar to MBP, but surprisingly the CNP probe was found clustered over the oligodendrocyte perikaryon. This perinuclear localization of CNP-mRNA has been demonstrated

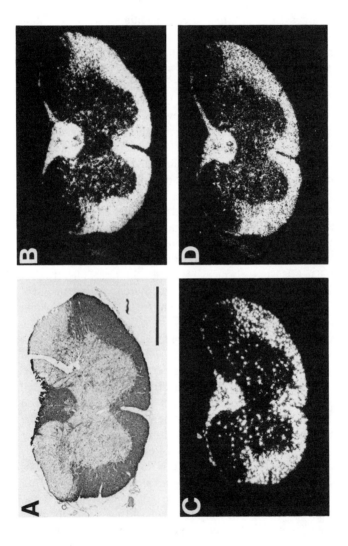

FIGURE 2. Expression of myelin-specific transcripts in 8-d postnatal mouse spinal cord. Cervical spinal cord cryostat sections, 8-d-old, were either immunochemically stained for CNP (A) or analyzed by *in situ* hybridization with probes specific for MBP exon 1 (B), PLP (C), MBP exon 2 (D), MAG (E), or CNP (F). All myelin-specific transcripts and the CNP protein were well expressed and limited to white matter regions. MBP transcripts were diffusely distributed while PLP, MAG, and CNP transcripts were clustered. (A) Brightfield illumination. (B—F) darkfield illumination. Autoradiographic exposures were 14 (B—E) or 40 d (F). Bar = 500 μm.

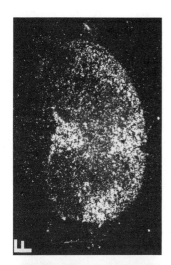

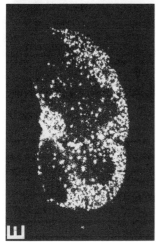

FIGURE 2 (continued).

by others.[18,20] Therefore, synthesis of a protein on free ribosomes does not imply that its mRNA will be found within processes of mature oligodendrocytes. Examination of neonatal CNS regions just beginning to myelinate revealed oligodendrocytes which had MBP-mRNA clustered over their cell body. This developmental shift in the localization of MBP-mRNAs has been noted in other *in situ* studies.[27,64] It is unclear if this developmental redistribution of MBP-mRNAs is somehow regulated or simply explained by young oligodendrocytes beginning to produce MBP-mRNA before they elaborate processes.

C. Remyelinating CNS Tissues

Armed with the basic knowledge of myelin-specific message expression during developmental myelination it now seemed feasible to examine remyelination following a demyelinating disease. Several rodent model systems exist in which remyelination occurs after a demyelinating event. We chose the mouse model of CNS demyelination induced by infection with the A-59 strain of mouse hepatitis virus (MHV-A59). This and other strains of MHV have been shown to infect and kill oligodendrocytes.[67-71] The majority of mice infected with MHV-A59 survive and rapidly clear the virus. Most of these mice develop myelin lesions of the spinal cord, which cause paresis and ataxia during the initial phase of disease (1 to 3 weeks). In the following weeks these mice spontaneously remyelinate and go on to a functional recovery.

Spinal cord sections were hybridized with a cRNA probe specific for viral sequences in order to determine the extent and location of viral replication in these tissues. Viral-mRNA was expressed maximally at 1 week postinfection (1 WPI) and was detected in both gray and white matter regions[21] (Figure 3). These transcripts were found in a pattern which suggested viral entry into the spinal cord at the dorsal and ventral root regions. The distribution of the virus was limited to white matter regions by 2 WPI, and after 4 WPI virus transcripts were no longer detectable. Immunostaining of adjacent sections for MBP demonstrated that this myelin protein was still normally distributed in white matter regions at 1 WPI. Staining of lipids with Sudan black also failed to demonstrate significant myelin loss at this time. Myelin lesions were easily demonstrable by MBP and lipid staining by 2 WPI and these lesions persisted past 6 WPI. These lesions were limited to white matter regions and almost always included either the dorsal or ventral root regions, the same areas where viral replication predominated.

In the initial phase of this study by Kristensson et al.,[72] spinal cord cryostat sections were probed *in situ* with small double stranded DNA fragments (from an MBP cDNA clone) which were 3' end labeled with ^{35}S-dATP. This probe mixture hybridized with all forms of MBP-specific transcripts. Analysis of spinal cord sections from demyelinated animals with this probe indicated reduced probe binding in lesioned areas at 2 WPI. These lesioned areas could be identified by the presence of inflammatory cell infiltrates visualized with a cresyl violet counterstain. MBP probe binding in some nonlesioned white matter

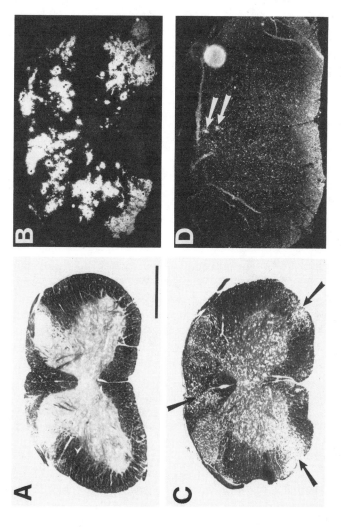

FIGURE 3. Expression of viral transcripts in spinal cords of MHV-A59 coronavirus infected mice. Cervical spinal cord sections from 1 WPI (A, B) or 4 WPI (C, D) animals were either stained with the lipidophilic dye Sudan black (A, C) or hybridized *in situ* with a cRNA probe for MHV-A59 transcripts (B, D; 8 day exposure). At 1 WPI, demyelinated lesions are not detected (A) while virus transcripts are highly expressed in both the gray and white matter regions. At 4 WPI, demyelinated lesions persist (C, arrows) even though expression of virus transcripts has been almost completely eliminated. Two foci of expression persist in this section (D, arrows). (A and C) Brightfield illumination. (B and D) Darkfield illumination. Bar = 500 μm.

regions increased in intensity at 3 to 4 WPI. Comparison of the grains from autoradiography with the location of lesions, characterized by inflammation and tissue degeneration, revealed that this increased MBP-mRNA signal was present at the edge of lesions as well as in the surrounding normal-appearing white matter. The demyelinated lesions were still devoid of MBP-specific probe binding. At 8 WPI reduced MBP-mRNA expression in lesions was no longer detectable while some white matter regions still had slightly increased levels of MBP-mRNA.

Analysis of the information gathered from this initial study generated the following questions:

1.   How does MHV-A59 replication relate to lesion formation?
2.   What happens to expression of myelin-specific genes other than MBP? Do they undergo increased expression in a similar pattern, with similar timing?
3.   Is MBP-mRNA expression during remyelination regulated by exon splicing as during developmental myelination?
4.   Is there a difference, either in timing or in transcript forms, between the increased MBP-mRNA expression within the demyelinated lesions as opposed to the increased MBP-mRNA expression outside the lesions?

We addressed these questions in subsequent studies in which additional genes and more time points were examined, especially at 1 WPI when lesions were not detectable and between 4 and 8 WPI when the lesions remyelinated. Spinal cord tissues were analyzed with probes for PLP-mRNA, MAG-mRNA, CNP-mRNA, and with probes specific for MBP exon 1 and 2 sequences. Probes used for this study were short single stranded oligonucleotides which could be designed to be exon specific, did not require a clone, and demonstrated very low nonspecific binding.[21,73]

In situ hybridization analysis for myelin-specific transcripts proved to be a more sensitive method for detection of lesion formation than histologic or immunologic staining since it revealed decreased mRNA expression even at 1 WPI (Figure 4). This decreased mRNA expression was demonstrable for MBP, PLP, CNP, and MAG, and was detected first in the dorsal and ventral root regions. The loss of myelin-specific gene transcripts was more extensive at 2 and 3 WPI, a time when cellular infiltrates and loss of myelin protein could also be demonstrated in the lesions. Thus, a sequential progression was demonstrated. First, the virus infects the spinal cord and replicates in selected cells, the oligodendrocytes appear to respond with reduced synthesis of myelin-specific transcripts, and finally, the animal develops paresis and ataxia at a time when myelin destruction is detectable.

When spinal cord sections from 4 to 6 WPI animals were examined, all myelin-specific gene probes were found to have increased binding within the demyelinated lesions. The pattern of binding for the individual probes was similar to that seen in normal tissue with PLP binding in high abundance and as tight clusters over cell bodies while MAG and CNP were less abundant and formed looser clusters. Even though all four myelin messages were increased at 6 WPI, the lesions were still demonstrable by decreased MBP immunostaining. decreased Sudan black staining, or the presence of cellular infiltrates (Figure 5). By 12 WPI these histological aspects as well as myelin gene message levels had returned to normal, the only indication of earlier demyelination was regions of axons which had thinner myelin sheaths when examined in epon-embedded sections.[21]

In an attempt to examine more specifically the process of remyelination in these mice, we utilized exon-specific oligonucleotide probes corresponding to MBP exon 1 and 2 information.[73] We knew from *in situ* hybridization analysis of developmental myelination that exon 2 information is well represented in MBP-mRNA early during myelination and that relatively few exon 2-containing MBP transcripts were present in adults. If MBP transcripts produced during remyelination followed this pattern of MBP expression, then the MBP exon 2 probe could specifically detect these transcript forms. The exon 1 probe was used for comparison since this exon is not spliced, and thus this probe will detect all MBP-mRNA forms.

When hybridized to 3 WPI spinal cord sections, the MBP exon 2 probe bound to discrete cell bodies within the demyelinated lesions. This clustered pattern of MBP probe binding is uncharacteristic of mature oligodendrocytes, but has been seen during early stages of developmental myelination.[8,27,64] These clusters of MBP exon 2 probe binding were easily detected against the very low level of diffuse exon 2 binding over the nonlesioned regions. Similar clusters were seen with the MBP exon 1 probe, but the higher binding of this probe over nonlesioned areas made its detection less obvious. In addition to the increased probe binding within the lesions, increased exon 1 probe binding was also detected in areas immediately adjacent to the lesions. The increased MBP transcript expression adjacent to lesions was not clustered and appeared to be composed predominantly of adult type transcripts since exon 2 probe binding was not appreciably increased in these areas.

After hybridization, selected spinal cord sections from different times postinfection were placed against X-ray film for quantitation. The X-ray images were digitized to measure optical density and then compared to autoradiographic standards. This quantitation indicated that MBP-mRNA expression continued to increase within the demyelinated lesions at 4, 5, and 6 WPI, but that the relative abundance of transcripts containing exon 2 information began to decline (Figure 6). This transient high expression of exon 2-containing MBP transcripts is an aspect of developmental myelination which appears to be reiterated during adult remyelination.

In immunocytochemical studies of rats demyelinated with cuprizone,[74] MBP has been detected in oligodendrocytes before remyelination. However, interpretation of immunostaining for MBP is complicated by staining of the preexisting

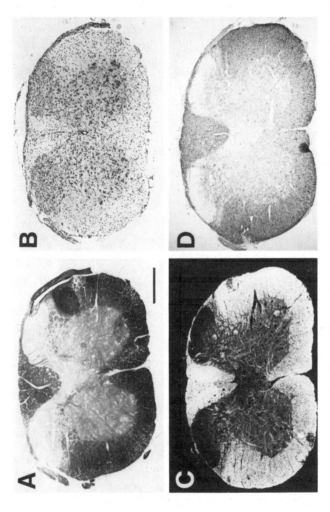

FIGURE 4. In coronavirus infected mice, loss of myelin-specific transcripts preceded demyelination. Cervical spinal cord sections from a 1 WPI mouse did not contain detectable demyelinating lesions either by Sudan black staining of lipids (A), or detectable cellular infiltrates by cresyl violet staining (B). Immunofluorescent staining for MBP (C) and immunohistochemical staining for CNP (D) also appeared normal. However, when neighboring sections were analyzed by *in situ* hybridization with probes for PLP (E), MBP (F), MAG (G), and CNP (H), identical regions of decreased transcript expression (arrows) were found for all four myelin-specific genes. The areas of transcript loss are the ventral root regions which contain high levels of viral transcripts at 1 WPI (see Figure 3B) and are one of the most common areas to become demyelinated later in the disease (see Figure 5A). (A, B, D) Brightfield illumination. (C) Fluorescence microscopy. (E—H) Darkfield illumination. Autoradiographic exposures were 10 (E, F) or 34 d (G, H). Bar = 500 μm.

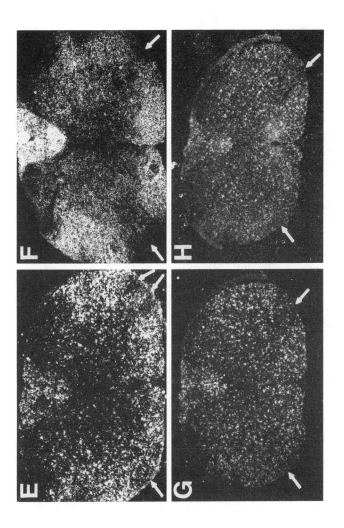

FIGURE 4 (continued).

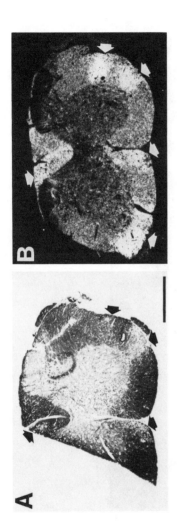

FIGURE 5. Expression of myelin gene transcripts were increased in demyelinated lesions during remyelination in C57B/6 mice. (A) A cervical spinal cord section from a 6 WPI mouse contains demyelinated lesions visualized by Sudan black staining (arrows). (B—F) *In situ* hybridization analysis of neighboring sections indicates that these same lesions express increased levels of the different myelin-specific genes (arrows). The probes used on the sections were specific for MBP exon 1, PLP, MBP exon 2, MAG, and CNP, respectively. Note the highly specific increase of MBP exon 2 signal in the lesioned area compared to the low signal over myelinated regions. (A) Brightfield illumination. (B—F) Darkfield illumination. Long autoradiographic exposures (75 d) were required for these emulsion dipped specimens because of isotope decay during previous film exposures. Bar = 500 μm.

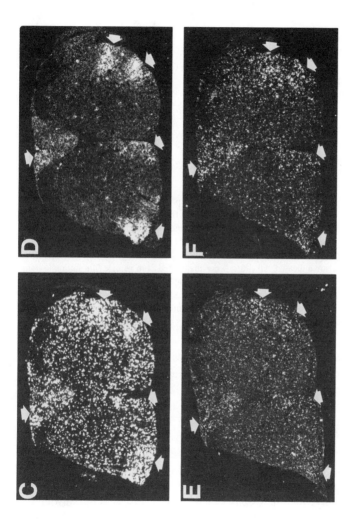

FIGURE 5 (continued).

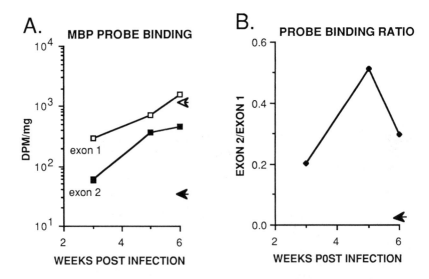

FIGURE 6.   Quantitative analysis of MBP exon specific probe binding. (A) The binding of MBP exon 1 probe (open squares) and MBP exon 2 probe (filled squares) was quantitated over demyelinated lesions from a film autoradiogram of 3, 5, and 6 WPI spinal cords. Points represent multiple readings from each of two sections. The standard error for all points was ≤10%. Values from an uninfected control animal (age matched to 6 WPI) for exon 1 (open arrowhead) and exon 2 (filled arrowhead) are indicated. All values were quantitated from a single film autoradiogram exposed for 15 d with all of the sections and autoradiographic standards. (B) The probe binding ratio (exon 2/exon 1) was calculated for the demyelinated regions (filled symbols) and the uninfected control (filled arrowhead). The relative abundance of MBP transcripts containing exon 2 information has peaked at 5 WPI and has sharply decreased at 6 WPI.

⁎

"old" myelin protein as well as macrophages laden with phagocytized myelin debris. In contrast, MBP-mRNA levels are quite low in normal adult white matter, thus allowing easier and more sensitive detection of newly synthesized MBP transcripts. Increased MBP-mRNA expression was detected at the edge of demyelinated lesions and seemed to radiate into the surrounding normal-appearing white matter. Based on the lack of exon 2 containing transcripts and the fact that this response occurs away from the demyelinated lesion, it is possible that surviving adult oligodendrocytes are being stimulated to produce increased numbers of transcripts. This could be explained if a signal was produced in these virus-induced demyelinating lesions which could diffuse into the surrounding tissue and stimulate oligodendrocytes and/or their progenitor cells present in adult CNS[75,76] to participate in myelin repair.[67] The possibility that such factor(s) might be produced by inflammatory cells in the lesions is supported by findings that mouse spleen cells[77,78] and human T lymphocytes[78] produce factors which promote proliferation and maturation of astrocytes and oligodendrocytes *in vitro*. Additionally, interleukin-2 has been shown to enhance MBP expression in cultured oligodendrocytes. Another potential source of stimulatory factors are type-1 astrocytes which have been shown to produce growth factors which act upon oligodendrocyte progenitors *in vitro*.[79,80]

This study indicates that *in situ* detection of gene transcripts is a sensitive and discriminating method to evaluate changes in gene expression. Subsets of transcripts can be assayed specifically, and because of rapid turnover, RNA transcripts probably reflect changes in gene expression more accurately and more specifically than proteins. Changes at the protein level occur more slowly and their detection is complicated by preexisting "old" protein which turns over more slowly than RNA transcripts. Our data and the data of others[81] indicate that myelin transcripts can be detected earlier than myelin proteins and that the presence of specific subsets of these transcripts indicates the earliest stage of myelination or remyelination.

IV. PROTOCOLS

A. Pretreatment of Cultured Cells

- Wash 2× in PBS to remove culture media
- Fix for 10 min with 4% paraformaldehyde in PBS at room temperature
- Wash 2× in PBS (cultures can be stored several days at 4°C)
- Permeabilize 3 min with 0.05% Triton X-100 in PBS
- Wash 2× in PBS
- Deproteinize for 10 min with 0.2 N HCl
- Wash 2× in PBS
- Further deproteinize for 15 min at 37° C with 1 to 10 µg/ml Proteinase K in 20 mM Tris pH 7.4 (predigest enzyme to destroy any RNase activity)
- Wash 2× in PBS containing 2 mg/ml glycine (to inactivate the enzyme) at room temperature
- Fix for 5 min with 4% paraformaldehyde in PBS
- Wash 2× in PBS
- Acetylate for 10 min at room temperature in freshly prepared 0.25% acetic anhydride in 0.1 M triethanolamine HCl/0.9% NaCl pH 8.0
- Dehydrate in graded ethanols (70, 80, 95, 100%) for 1 to 2 min each
- Delipidate for 5 min in chloroform (this step can often be omitted for tissue culture cells and must be avoided if cultures are in plastic dishes)
- 100% ethanol
- 95% ethanol
- Air dry in a vertical position and avoid dust (specimens are now ready for the hybridization step)

B. Pretreatment of Tissue Sections

Sections which were cut from frozen tissue on a cryostat, thaw mounted onto 1% gelatin coated slides, and stored at –20° C are removed from the freezer when needed and treated in the following manner:

- Place slides on aluminum foil for 10 min at room temperature
- Fix for 5 min with 4% paraformaldehyde in PBS
- Wash 2× in PBS
- Acetylate for 10 min at room temperature in freshly prepared 0.25% acetic anhydride in 0.1 M triethanolamine HCI/0.9% NaCI pH 8.0
- Dehydrate in graded ethanols (70, 80, 95, 100%) for 1 to 2 min each
- Delipidate for 5 min in chloroform (this step is particularly useful for nervous tissue which contains lipid-rich myelin)
- 100% ethanol
- 95% ethanol
- Air dry in a vertical position and avoid dust (specimens are now ready for the hybridization step)

C. Production of Single Stranded ^{35}S-Labeled Probes

The DNA sequence to be used for probe production must be cloned into a suitable *in vitro* transcription vector which contains one or more of the RNA polymerase binding sites (SP6, T7, T3). The ideal construct has the probe sequence located between two different polymerase binding sites so that sense and antisense RNA probes can be produced from the same construct.

- *In vitro* transcription mixture:

X μl	Add DEPC-treated water to bring final volume to 15 μl
3.0 μl	5× buffer (200 mM Tris-HCI, pH 7.5; 30 mM MgCl$_2$; 10 mM spermadine; 50 mM NaCl)
0.75 μl	Dithiothreitol (DTT, 100 mM)
1.0 μl	RNase inhibitor (33 U/μl)
0.75 μl	GTP (10 mM)
0.75 μl	CTP (10 mM)
0.75 μl	ATP (10 mM)
0.75 μl	UTP (0.167 mM)
2.5 μl	[α ^{35}S]-UTP (>1000 Ci/mmol; 10 Ci/ml)
2.0 μl	DNA (0.5 μg/μl) linearized
X μl	RNA polymerase (10 to 30 U; enzyme is determined by the binding site contained in the DNA)
———	
15 μl	Total volume

- Incubate 60 to 90 min at 37°C
- Add 2 U of DNase I in 40 to 60 μl of 40 mM Tris, 10 mM MgCl$_2$ pH 7.5 made with DEPC treated water; incubate at 37°C for 15 min to digest template DNA

- Extract with phenol-chloroform-isoamyl alcohol (50:49:1)
- Separate probe RNA from unincorporated nucleotides on a G-50 Sephadex column
- Pool fractions
- Ethanol precipitate using $\frac{1}{10}$ vol of 3 M sodium acetate and 2.5 vol of ethanol
- Resuspend in 150 µl of DEPC treated water
- Add 150 µl of 2× hydrolysis buffer (80 mM NaHCO$_3$, 120 mM Na$_2$CO$_3$, pH 10.2)
- Incubate at 60°C for t min, where:

t	=	Lo - Lf/k Lo Lf[25]
t	=	time in minutes
Lo	=	original length in kb
Lf	=	final fragment length in kb
k	=	rate constant of 0.11/kb min

- Stop hydrolysis with addition of sodium acetate pH 6 to a concentration of 0.1 M and glacial acetic acid to 0.5%
- Ethanol precipitate
- Resuspend in 50 µl of TE (10 mM Tris, 1 mM EDTA) containing 100 mM DTT, and 60 U RNase inhibitor
- Count 1 µl
- Probe can be stored at 4°C for 1 to 2 weeks or at −70°C for slightly longer periods, however, freshly prepared probe is recommended

D. Production of ³⁵S End-Labeled Oligonucleotide Probes

The enzyme terminal deoxynucleotidyl transferase (TdT) is used to add multiple ^{35}S-dATP molecules to the 3′-end of synthetic oligonucleotides. Oligonucleotides are purified by polyacrylamide gel electrophoresis and used at a concentration of 200 ng/µl.

- Labeling reaction mixture:

X µl	Add dH$_2$O to bring final volume to 50 µl
25 µl	0.2 M potassium cacodylate pH 7.2
1 µl	BSA (10 mg/ml)
2 µl	50 mM cobalt chloride (prepared every 2 weeks)
8 µl	[α ^{35}S]-dATP (>1000 Ci/mmol; 10 Ci/ml)
1 µl	DNA (200 ng/µl)
X µl	TdT (50 to 70 U)
50 µl	Total volume

- Incubate at 37° C for 2 h
- Add 100 µl of TE and 4 µl of tRNA (20 µg/ml)
- Extract with phenol-chloroform-isoamyl alcohol (50:49:1)
- Ethanol precipitate
- Resuspend in TE containing 100 µ*M* DTT
- Count 1 µl and store remainder at 4°C

E. Hybridization Conditions

The hybridization buffer is made in advance, aliquoted, and stored at –20°C. It contains:

> 4× SSC
> 50% deionized formamide
> 1× Denhardt's
> 500 µg/ml sheared single strand salmon sperm DNA
> 250 µg/ml yeast tRNA
> 10% dextran sulfate (mol wt = 500,000)

- Warm the hybridization buffer and add 1 µl of 5 *M* DTT and 1 to 2×10^6 cpm of probe for each 50 µl of mixture
- Cover the dried cells or tissue sections with the complete hybridization mixture (~20 µl/cm²)
- Cover with a correctly sized piece of parafilm and incubate in a humid chamber at 37°C overnight

F. Posthybridization Washing for Oligonucleotide Probes (48 mers)

- 1× SSC, to remove parafilm covering sections
- 1× SSC, 3 quick rinses
- 2× SSC with 50% formamide at 40°C, 4 changes of 15 min each
- 1× SSC at room temperature, 2 changes of 30 min each
- dH$_2$O, a brief dip
- 70% ethanol
- Air dry

G. Posthybridization Washing for cRNA Probes

- 1× SSC, to remove parafilm covering sections
- 1× SSC, 3 quick rinses
  RNase digestion with RNase A (100 µg/ml) and RNase T1 (10 U/ml) in 10 m*M* Tris pH 8.0, 1 m*M* EDTA, 0.5 *M* NaCl for 30 min at 37°C
- 1× SSC, 2 changes
- 2× SSC with 50% formamide at 40°C, 2 changes of 15 min each
- 0.1× SSC at 45 to 55°C, 2 changes of 15 min each

- dH$_2$O, a brief dip
- 70% ethanol
- Air dry

H. Autoradiographic Detection

After air drying, the hybridized slides may be immobilized in a film cassette and directly apposed to X-ray film (Kodak®X-OmatAR). If one is careful not to damage the tissue sections, they may be used for multiple exposures. Unlike emulsion dipped autoradiograms, film exposures yield a linear response between radioactivity and film density. By exposing many slides and radioactive standards on the same film one can obtain quantitative data on probe binding based on film density measurements. Film exposures are also a good way to estimate exposure times for emulsion dipping.

Emulsion dipped slides can be combined with histologic or immunologic stains in order to gain additional information about the cells expressing the mRNA of interest.

- Dilute emulsion (Kodak NTB2) 1:1 in H$_2$O and warm to 45°C under appropriate safelight conditions
- Dip slides, air dry, expose dessicated at 4°C
- Develop slides with D-19 (Kodak, $^1/_2$ strength) at 15°C for 4 min
- Wash 1 min in H$_2$O
- 5 min in Kodak fixer
- Wash at least 5 min in H$_2$O
- Counterstain as desired and coverslip

ACKNOWLEDGMENTS

This work was performed in the Laboratory of Molecular Genetics and the Laboratory of Viral and Molecular Pathogenesis, National Institute of Neurological Disorders and Stroke. I would like to thank all the members of these laboratories for their help and support, most notably Drs. M. Dubois-Dalcq, V. Friedrich, Jr., N. Zeller, and K. Kristensson. I am also grateful for the photographic expertise of Mr. R. Rusten.

REFERENCES

1. **Cohen-Hagaenauer, O., Barton, P. J., Van Cong, N., Cohen, A., Masset, A., Bucking- ham, M., and Frezal, J.,** Chromosomal assignment of two myosin alkali light-chain genes encoding the ventricular/slow skeletal muscle isoform and the atrial/fetal muscle isoform (MYL3, MYL4), *Hum. Genet.,* 81, 278, 1989.
2. **Capco, D. G. and Jeffrey, W. R.,** Differential distribution of poly(A)-containing RNA in the embryonic cells of *Oncopeltus fasciatus, Dev. Biol.,* 67, 137, 1978.
3. **Smith, G. H., Doherty, P. J., Stead, R. B., Gorman, C. M., Graham, D. E., and Howard, B. H.,** Detection of transcription and translation *in situ* with biotinylated molecular probes in cells transfected with recombinant DNA plasmids, *Anal. Biochem.,* 156, 17, 1986.
4. **Angerer, L. M. and Angerer, R. C.,** Detection of poly A^+ RNA in sea urchin eggs and embryos by quantitative *in situ* hybridization, *Nucl. Acids Res.,* 9, 2819, 1981.
5. **Tremblay, Y., Tretjakoff, I., Peterson, A., Antakly, T., Zhang, C. X., and Drouin, J.,** Pituitary-specific expression and glucocorticoid regulation of a proopiomelanocortin fusion gene in transgenic mice, *Proc. Natl. Acad. Sci. U.S.A.,* 85, 8890, 1988.
6. **Marks, J. R., Lin, J., Hinds, P., Miller, D., and Levine, A. J.,** Cellular gene expression in papillomas of the choroid plexus from transgenic mice that express the simian virus 40 large T antigen, *J. Virol.,* 63, 790, 1989.
7. **Zeller, N. K., Behar, T. N., Dubois-Dalcq, M. E., and Lazzarini, R. A.,** The timely expression of myelin basic protein gene in cultured rat brain oligodendrocytes is independent of continuous neuronal influences, *J. Neurosci.,* 5, 2955, 1985.
8. **Jordan, C., Friedrich, V., Jr., Dubois-Dalcq, M.,** *In situ* hybridization analysis of myelin gene transcripts in developing mouse spinal cord, *J. Neurosci.,* 9, 248, 1989.
9. **Gendelman, H. E., Koenig, S., Aksamit, A., and Venkatesan, S.,** *In situ* hybridization for detection of viral nucleic acid in cell cultures and tissues, in *In Situ Hybridization In Brain,* Uhl, G. R., Ed., Plenum Press, New York, 1986, 203.
10. **Sideras, P., Funa, K., Zalcberg-Quintana, I., Xanthopoulos, K. G., Kisielow, P., and Palacios, R.,** Analysis by *in situ* hybridization of cells expressing mRNA for interleukin 4 in the developing thymus and in peripheral lymphocytes from mice, *Proc. Natl. Acad. Sci. U.S.A.,* 85, 218, 1988.
11. **Dony, C. and Gruss, P.,** Expression of a murine homeobox gene precedes the induction of c-fos during mesodermal differentiation of P19 teratocarcinoma cells, *Differentiation,* 37, 115, 1988.
12. **Schechter, H., Holtzclaw, L., Sadiq, F., Kahn, A., and Devaskar, S.,** Insulin synthesis by isolated rabbit neurons, *Endocrinology,* 123, 505, 1988.
13. **Harper, M. E. and Marselle, L. M.,** RNA detection and localization in cells and tissue sections by *in situ* hybridization of ^{35}S-labeled RNA, *Methods Enzymol.,* 151, 539, 1987.
14. **Holmes, E., Hermanson, G., Cole, R., and deVellis, J.,** Developmental expression of glial-specific mRNAs in primary cultures of rat brain visualized by *in situ* hybridization, *J. Neurosci. Res.,* 19, 389, 1988.
15. **Singer, R. H., Lawrence, J. B., and Villnave, C.,** Optimization of *in situ* hybridization using isotopic and non-isotopic detection methods, *Biotechniques,* 4, 230, 1986.
16. **Bandtlow, C. W., Heumann, R., Schwab, M. E., and Thoenen, H.,** Cellular localization of nerve growth factor synthesis by *in situ* hybridization, *EMBO J.,* 6, 891, 1987.
17. **Kristensson, K., Zeller, N. K., Dubois-Dalcq, M. E., and Lazzarini, R. A.,** Expression of myelin basic protein gene in the developing rat brain as revealed by *in situ* hybridization, *J. Histochem. Cytochem.,* 34, 467, 1986.
18. **Vogel, U. S., Reynolds, R., Thompson, R. J., and Wilkin, G. P.,** Expression of the 2′,3′-cyclic nucleotide 3′-phosphohydrolase gene and immunoreactive protein in oligodendrocytes as revealed by *in situ* hybridization and immunofluorescence, *Glia,* 1, 184, 1988.

19. **Young, W. S.,** *In situ* hybridization histochemical detection of neuropeptide mRNAs using DNA and RNA probes, *Methods Enzymol.,* 168, 702, 1989.

20. **Trapp, B. D., Bernier, L., Andrews, S. B., and Colman, D. R.,** Cellular and subcellular distribution of 2′,3′-cyclic nucleotide 3′-phosphodiesterase and its mRNA in the rat central nervous system, *J. Neurochem.,* 51, 859, 1988.

21. **Jordan, C. A., Friedrich, V. L., Jr., Godfraind, C., Cardellechio, C. B., Holmes, K. V., and Dubois-Dalcq, M.,** Expression of viral and myelin gene transcripts in a murine CNS demyelinating disease caused by a coronavirus, *Glia,* 2, 318, 1989.

22. **Trapp, B. D. and Quarles, R. H.,** Immunocytochemical localization of the myelin-associated glycoprotein. Fact or artifact?, *J. Neuroimmunol.,* 6. 231, 1984.

23. **Webster, H. de F., Lamperth, L., Favilla, J. T., Lemke, G., Tesin, D., and Manuelidis, L.,** Use of a biotinylated probe and *in situ* hybridization for light and electron microscopic localization of P$_o$ mRNA in myelin-forming Schwann cells, *Histochemistry,* 86, 441, 1987.

24. **Uhl, G. R., Ed.,** *In Situ Hybridization in Brain,* Plenum Press, New York, 1986, 267.

25. **Cox, K. H., DeLeon, D. V., Angerer, L. M., and Angerer, R. C.,** Detection of mRNAs in sea urchin embryos by *in situ* hybridization using asymmetric RNA probes, *Dev. Biol.,* 101, 485, 1984.

26. **Smith, G. H.,** *In situ* detection of transcription in transfected cells using biotin-labeled molecular probes, *Methods Enzymol.,* 151, 530, 1987.

27. **Trapp, B. D., Moeuch, T., Pulley, M., Barbosa, E., Tennekoon, G., and Griffith, J.,** Spatial segregation of mRNA encoding myelin-specific proteins, *Proc. Natl. Acad. Sci. U.S.A.,* 84, 7773, 1987.

28. **Lawrence, J. B. and Singer, R. H.,** Quantitative analysis of *in situ* hybridization methods for the detection of actin gene expression, *Nucl. Acids Res.,* 13, 1777, 1985.

29. **Heumann, R., Korsching, S., Bandtlow, C., and Thoenen, H.,** Changes of nerve growth factor synthesis in nonneuronal cells in response to sciatic nerve transection, *J. Cell Biol.,* 104, 1623, 1987.

30. **Langer, P. R., Waldrop, A. A., and Ward, D. C.,** Enzymatic synthesis of biotin-labeled polynucleotides: novel nucleic acid affinity probes, *Proc. Natl. Acad. Sci. U.S.A.,* 78, 6633, 1981.

31. **Dooley, S., Radtke, J., Blin, N., and Unteregger, G.,** Rapid detection of DNA-binding factors using protein blotting and digoxigenin-dUTP marked probes, *Nucl. Acids Res.,* 16, 11839, 1988.

32. **Benjamins, J. A. and Smith, M. E.,** Metabolism of myelin, in *Myelin,* Morell, P., Ed., Plenum Press, New York, 1984, 225.

33. **Raine, C. S.,** Biology of disease: the analysis of autoimmune demyelination: its impact on multiple sclerosis, *Lab. Invest.,* 50, 608, 1984.

34. **Silberberg, D. H.,** Pathogenesis of demyelination, in *Multiple Sclerosis,* McDonald, W. I., and Silberberg, D. H., Eds., Butterworths, London, 1986, 99.

35. **Lees, M. B. and Brostoff, S. W.,** Proteins of myelin, in *Myelin,* Morell, P., Ed., Plenum Press, New York, 1984, 197.

36. **Norton, W. T. and Cammer, W.,** Isolation and characterization of myelin, in *Myelin,* Morell, P., Ed., Plenum Press, New York, 1984, 147.

37. **Duncan, I. D., Hammang, J. P., and Jackson, K. F.,** Jimpy myelin lacks PLP and has a defect in the intraperiod line, *Soc. Neurosci. Abstr.,* 13, 118, 1987.

38. **Privat, A. C., Jacque, J. M., Bourre, P., Doupouey, P., and Baumann, N.,** Absence of the major dense line in myelin of the mutant mouse "shiverer", *Neurosci. Lett.,* 12, 107, 1979.

39. **Omlin, F. X., Webster, H. de F., Paklovitz, C. G., and Cohen, S. R.,** Immunocytochemical localization of basic protein in major dense line regions of central and peripheral myelin, *J. Cell Biol.,* 95, 242, 1982.

40. **Trapp, B. P., Bernier, L., Andrews, S. B., and Colman, D. R.,** Cellular and subcellular distribution of 2',3'-cyclic nucleotide 3'-phosphodiesterase and its mRNA in the rat central nervous system, *J. Neurochem.,* 51, 859, 1988.

41. **Braun, P. E., Sandillon, F., Edwards, A., Matthieu, J. M., and Privat, A.,** Immunocytochemical localization by electron microscopy of 2',3'-cyclic nucleotide 3'-phosphodiesterase in developing oligodendrocytes of normal and mutant brain, *J. Neurosci.,* 8, 3057, 1988.

42. **Williams, A. F.,** A year in the life of the immunoglobulin superfamily, *Immunol. Today,* 8, 298, 1987.

43. **Quarles, R. H.,** Myelin associated glycoprotein in development and disease, *Dev. Neurosci.,* 6, 285, 1985.

44. **Sternberger, N. H., Quarles, R. H., Itoyama, Y., and Webster, H. de F.,** Myelin-associated glycoprotein demonstrated immunocytochemically in myelin and myelin-forming cells of devloping rat, *Proc. Natl. Acad. Sci. U.S.A.,* 76, 1510, 1979.

45. **Roussel, G. and Nussbaum, J. L.,** Comparative localization of Wolfgram W1 and myelin basic protein in the rat brain during ontogenesis, *Histochem. J.,* 13, 1029, 1981.

46. **Hartman, B. K., Agrawal, N. C., Agrawal, D., and Kalmback, S.,** Development and maturation of central nervous system myelin: comparison of immunohistochemical localization of proteolipid protein and basic protein in myelin in oligodendrocytes, *Proc. Natl. Acad. Sci. U.S.A.,* 79, 4217, 1982.

47. **Monge, M., Kadiisky, D., Jacque, C., and Zalc, B.,** Oligodendroglial expression and deposition of four major myelin constituents in the myelin sheath during development: an *in vivo* study, *Dev. Neurosci.,* 8, 222, 1986.

48. **Zeller, N. K., Hunkeler, M. J., Campagnoni, A. T., Sprague, J., and Lazzarini, R. A.,** Characterization of mouse myelin basic protein messenger, RNAs with a myelin basic protein cDNA clone, *Proc. Natl. Acad. Sci. U.S.A.,* 81, 18, 1984.

49. **de Ferra, F., Engh, H., Hudson, L., Kamholz, J., Puckett, C., Molineaux, S., and Lazzarini, R. A.,** Alternative splicing accounts for the four forms of myelin basic protein, *Cell,* 43, 721, 1985.

50. **Naismith, A. L., Hoffman-Chudzik, E., Tsui, L.-C., and Riordan, J. R.,** Study of the expression of myelin proteolipid protein (lipophilin) using a cloned complementary DNA, *Nucl. Acids Res.,* 13, 7413, 1985.

51. **Takahashi, N., Roach, A., Teplow, D. B., Prusiner, S. B., and Hood, L.,** Cloning and characterization of the myelin basic protein gene from mouse: one gene can encode both 14 kd and 18.5 kd MBPs by alternate use of exons, *Cell,* 42, 139, 1985.

52. **Diehl, H.-J., Schaich, M., Budzinski, R.-M., and Stoffel, W.,** Individual exons encode the integral membrane domains of human myelin proteolipid protein, *Proc. Natl. Acad. Sci. U.S.A.,* 83, 9807, 1986.

53. **Arquint, M., Roder, J., Chia, L.-S., Down, J., Wilkinson, D., Bayley, H., Braun, P., and Dunn, R.,** Molecular cloning and primary structure of myelin-associated glycoprotein, *Proc. Natl. Acad. Sci. U.S.A.,* 84, 600, 1987.

54. **Bernier, L., Alverez, F., Norgard, E. M., Raible, D. W., Mentaberry, A., Schembri, J. G., Sabatini, D., D., and Colman, D. R.,** Molecular cloning of a 2',3'-cyclic nucleotide 3'-phosphodiesterase: mRNAs with different 5' ends encode the same set of proteins in nervous and lymphoid tissues, *J. Neurosci.,* 7, 2703, 1987.

55. **Lai, C., Brow, M. A., Naxe, K.-A., Noronha, A. B., Quarles, R. H., Bloom, F. E., Milner, R. J., and Sutcliffe, J. G.,** Two forms of 1B236/myelin-associated glycoprotein, a cell adhesion molecule for postnatal neural development, are produced by alternative splicing, *Proc. Natl. Acad. Sci. U.S.A.,* 84, 4337, 1987.

56. **Hudson, L. D., Berndt, J., Puckett, C., Kozak, C. A., and Lazzarini, R. A.,** Aberrant splicing of proteolipid protein mRNA in the dysmyelinating jimpy mutant mouse, *Proc. Natl. Acad. Sci. U.S.A.,* 84, 1454, 1987.

57. **Barbarese, E. and Pfeiffer, S. E.,** Developmental regulation of myelin basic protein in dispersed cultures, *Proc. Natl. Acad. Sci. U.S.A.,* 78, 1953, 1981.

58. **Campagnoni, C. W., Carey, G. D., and Campagnoni, A. T.,** Synthesis of myelin basic proteins in the developing mouse brain, *Arch. Biochem. Biophys.,* 190, 143, 1978.

59. **Sternberger, N. H., Itoyama, Y., Kies, M., and Webster, H. de F.,** Myelin basic protein demonstrated immunocytochemically in oligodendroglia prior to myelin sheath formation, *Proc. Natl. Acad. Sci. U.S.A.,* 75, 2521, 1978.

60. **Mentaberry, A., Adesnik, M., Atchison, M., Norgard, E. M., Alvarez, F., Sabatini, D. D., and Colman, D. R.,** Small basic proteins of myelin from central and peripheral nervous systems are encoded by the same gene, *Proc. Natl. Acad. Sci. U.S.A.,* 83, 1111, 1986.

61. **Carson, J. H., Nielson, M. L., and Barbarese, E.,** Developmental regulation of myelin basic protein expression in the mouse brain, *Dev. Biol.,* 96, 485, 1983.

62. **Kamholz, J., de Ferra, F., Puckett, C., and Lazzarini, R.,** Identification of three forms of human myelin basic protein by cDNA cloning, *Proc. Natl. Acad. Sci. U.S.A.,* 83, 4962, 1986.

63. **Small, R. K., Riddle, P., and Noble, M.,** Evidence for migration of oligodendrocyte-type-2 astrocyte progenitor cells into the developiong rat optic nerve, *Nature,* 328, 155, 1987.

64. **Verity, A. N. and Campagnoni, A. T.,** Regional expression of myelin protein genes in the developing mouse brain: *in situ* hybridization studies, *J. Neurosci. Res.,* 21, 238, 1988.

65. **Lamperth, L., Manuelidis, L., and Webster, H. de F.,** Non myelin-forming perineuronal Schwann cells in rat trigeminal ganglia express P_o myelin glycoprotein mRNA during postnatal development, *Mol. Res.,* 5, 177, 1989.

66. **Karin, N. J. and Waehneldt, T. V.,** Biosynthesis and insertion of Wolfgram protein into optic nerve membranes, *Neurochem. Res.,* 10, 897, 1985.

67. **Herndon, R. M., Price, D. L., and Weiner, L. P.,** Regeneration of oligodendroglia during recovery from demyelinating disease, *Science,* 195, 693, 1977.

68. **Knobler, R. L., Dubois-Dalcq, M., Haspel, M. V., Claysmith, A. P., Lampert, P. W., and Oldstone, M. B. A.,** Selective localization of wild type and mutant mouse hepatitis virus (JHM strain) antigens in CNS tissue by fluorescence, light and electron microscopy, *J. Neuroimmunol.,* 1, 81, 1981.

69. **Sorensen, O., Coulter-Mackie, M., Percy, D., and Dales, S.,** *In vivo* and *in vitro* models of demyelinating diseases, *Adv. Exp. Med. Biol.,* 142, 271, 1981.

70. **Lavi, E., Gilden, D. H., Highkin, M. K., and Weiss, S. R.,** Persistence of mouse hepatitis virus A59 RNA in a slow virus demyelinating infection in mice as detected by *in situ* hybridization, *J. Virol.,* 51, 563, 1984.

71. **Woyciechowska, J. L., Trapp, B. D., Patrick, D. H., Shekarchi, I. C., Leinikki, P. O., Sever, J. L., and Holmes, K. V.,** Acute and subacute demyelination induced by mouse hepatitis virus strain A59 in C3H mice, *J. Exp. Pathol.,* 1, 295, 1984.

72. **Kristensson, K., Holmes, K. V., Duchala, C. S., Zeller, N. K., Lazzarini, R. A., and Dubois-Dalcq, M.,** Increased levels of myelin basic protein transcripts in virus-induced demyelination, *Nature,* 322, 544, 1986.

73. **Jordan, C. A., Friedrich, V. L., Jr., de Ferra, F., Weismiller, D. G., Holmes, K. V., and Dubois-Dalcq, M.,** Differential exon expression in myelin basic protein transcripts during CNS remyelination, *Cell. Mol. Neurobiol.,* 10, 3, 1990.

74. **Ludwin, S. and Sternberger, N.,** An immunohistochemical study of myelin proteins during remyelination in the central nervous system, *Acta Neuropathol.,* 63, 240, 1984.

75. **ffrench-Constant, C. and Raff, M. C.,** Proliferating bipotential glial progenitor cells in adult rat optic nerve, *Nature,* 319, 499, 1986.

76. **Wolswijk, G. and Noble, M.,** Identification of an adult-specific glial progenitor cell, *Development,* 105, 387, 1989.

77. **Fontana, A., Dubs, R., Merchant, R., Balsinger, S., and Grob, P. J.,** Glial cell stimulating factor (GSF): a new lymphokine. I. Cellular sources and partial purification of murine GSF, role of cytoskeleton and protein synthesis in this production, *J. Neuroimmunol..* 2, 55, 1981.

78. **Fontana, A., Otz, U., DeWeck, A. L., and Grob, P. J.,** Glial cell stimulating factor (GSF): a new lymphokine. II. Cellular sources and partial purification of human GSF, *J. Neuroimmunol.,* 2, 73, 1981.

79. **Hughes, S. M., Lillien, L. E., Raff, M. C., Rohrer, H., and Sendtner, M.,** Ciliary neurotrophic factor induces type-2 astrocyte differentiation in culture, *Nature,* 335, 70, 1988.

80. **Raff, M. C., Lillien, L. E., Richardson, W. D., Burne, J. F., and Nobel, M. D.,** Platelet-derived growth factor from astrocytes drives the clock that times oligodendrocyte development in culture, *Nature,* 333, 562, 1988.

81. **De Vitry, F., Gomes, D., Rataboul, P., Dumas, S., Hillion, J., Catelon, J., Delaunoy, J.-P., Tixier-Vidal, A., and Dupouey, P.,** Expression of carbonic anhydrase II gene in early brain cells as revealed by *in situ* hybridization and immunohistochemistry, *J. Neurosci. Res.,* 22, 120, 1989.

Chapter 4

THE USE OF *IN SITU* HYBRIDIZATION TO STUDY GENE EXPRESSION IN MOUSE NEUROLOGICAL MUTANTS*

Gretchen D. Frantz and Allan J. Tobin

TABLE OF CONTENTS

*Reprint requests should be addressed to Dr. Allan Tobin, Department of Biology, UCLA, Los Angeles, CA 90024-1606.

I. INTRODUCTION

Abnormalities in gene expression are likely to underlie some of the behavioral changes of inherited neurological and neuropsychiatric diseases. Anatomical examination of the brains of affected patients or experimental animals often, but not always, reveals changes in specific cells (e.g., striatal medium spiny neurons in Huntington's disease) or in the organization of specific groups of cells (e.g., cerebellar Purkinje cells in *reeler* mice). Comparisons of the patterns of gene expression in specific cell populations in affected and unaffected brains can reveal much about the influence of altered environments on the expression of specific genes.

In many cases, the severely altered phenotypes result from single mutations which may influence the expression of many genes. Some mutations, such as those in cis-regulatory elements, may change the transcription rate of a single gene within a cell. Other mutations, such as those in genes encoding trans-acting regulatory factors, may affect the expression of many genes in a single cell.

Mutations may also have less direct effects. By changing the properties of one cell, a mutation also can alter the environment of neighboring cells. Within the nervous system the situation may be still more complex. The phenotype of a particular neuron may not only affect its immediate physical neighbors, but may also influence neurons in distant regions that form connections with the deviant cell.

In this chapter, we review the use of *in situ* hybridization to examine gene expression in neurological mutants of mice. We summarize the contributions of these studies to the knowledge of the extrinsic and intrinsic influences on gene expression in neural cells.

A. Extrinsic vs. Intrinsic Gene Regulation

Researchers have experimentally altered the environment and synaptic connections of neurons in a variety of ways. Some strategies involve the modification or elimination of cell populations providing afferent inputs or efferent targets to specific cell groups. These approaches include: (1) elimination of cell populations by X-irradiation of dividing cells,[1] (2) transplantation of cell populations into abnormal environments,[2-4] (3) elimination of specific afferent connections with stereotactic lesions,[5,6] (4) observation of cell behavior in culture,[7-9] (5) creation of genetically mosaic chimeras,[10-13] (6) generation of transgenic mice,[14,15] and (7) the examination of mouse neurological mutants.[16]

These studies have revealed that some cells can function autonomously in abnormal environments, while others function inefficiently, and some even die. The published reports, however, have not yielded a clear answer to the question of why some cells survive and others die.

The affected cells may contain a set of genes whose expression is altered qualitatively or quantitatively in response to the altered environment. Likewise, the affected cells may contain a set of genes whose expression is rigidly governed by an intrinsic program established early during development. *In situ* hybridization provides a means for studying which genes respond to abnormal environments. Such experiments may lead to an understanding of processes that regulate gene expression, both during normal development and in pathological conditions.

B. Mouse Neurological Mutants

The genetically reproducible behavioral abnormalities (seizures, ataxias) demonstrated by neurologically mutant mice led to their identification and classification.[16] Examination of the brains of mouse neurological mutants often reveals some cells that appear to be normal but are nonetheless placed in an abnormal milieu, due to the effect of the mutant gene upon its primary targets. The abnormal environment results from reproducible aberrations that include neuronal degeneration, axonal overgrowth, and neuronal malpositioning. Each mutation may affect one or more distinct developmental stages. For example, several well-studied cerebellar mutations (*reeler, weaver,* and *staggerer*) affect neuronal proliferation, migration and/or synaptogenesis.[17]

Inbred mouse strains provide some specific advantages to the study of neurological disease. First, the disease phenotype is highly reproducible in many animals, so a constant source of mutants exists. Second, the effects of single gene mutations can be studied in isolation by comparing mutants to unaffected littermates which are homozygous at every other allele. Third, the interaction of the disease gene with the background genotype can be studied by outcrossing.

C. Alternate Approaches to the Study of Altered Gene Expression: Advantages and Limitations

Many studies have taken advantage of the reproducible defects within mouse mutants to identify gene products whose expression is altered in particular brain regions. Biochemical studies have demonstrated overall changes in enzymes, antigens, and receptors. In these studies, however, the affected cells usually cannot be identified.

Autoradiography with labeled receptor ligands has demonstrated altered expression of particular receptor types. Since many receptors are located on neuronal processes, however, the identities of affected cells are often uncertain.

Histochemistry and immunohistochemistry have provided more accurate assessments of the cell types with altered expression of gene products. These techniques have revealed altered expression in some mutants. In most cases, however, the data are qualitative, rather than quantitative.

Table 1 summarizes the published studies that document altered gene products in mouse neurological mutants using the approaches described above. Below, we compare these techniques and the kind of information they provide about gene expression in mouse neurological mutants.

1. Level of Regulation

Immunohistochemical, histochemical, and ligand binding techniques can detect changes in the levels of functional proteins. They cannot, however, distinguish between changes in the expression of the particular gene, on the one hand, and later changes, e.g., post-translational modification, on the other. For example, the conversion of the embryonic to the adult form of the neural cell adhesion molecule (N-CAM) is delayed in *staggerer* mutants, but this delay apparently results from altered post-translational sialylation, rather than from altered transcription of the N-CAM gene.[18]

In situ hybridization measures mRNA levels and thus reveals only pretranslational changes. Such studies should lead to a sharper definition of the informational level at which regulation occurs. For example, mutations affecting promoters or transacting factors would alter transcription rates or developmental expression. Mutations in untranslated regions could alter mRNA stability or intracellular distribution. Although RNA blots also can provide data concerning transcription rates and mRNA stability, *in situ* hybridization has the advantage over RNA blots of revealing whether altered gene expression occurs in a select subset of cells within a region.

2. Cellular Identification

Immunohistochemical studies suffer from the limitation that the cell in which a particular compound is measured may not be the same cell that produces it. Nerve growth factor (NGF), for example, is synthesized by cells in the hippocampus or neocortex, then taken up and retrogradely transported by a second set of cells dependent upon NGF for their survival (e.g., neurons of the septal nuclei, nucleus of the diagonal band of Broca, and nucleus basalis of Meynert).[19]

The cerebella of *Purkinje cell degeneration (pcd)* mutant mice contain elevated levels of NGF, based upon NGF radioimmunoassay data.[20] *In situ* hybridization studies could help localize the cellular source of the elevated levels of NGF in the *pcd* cerebella.

Already, *in situ* hybridization studies have demonstrated NGF mRNA in the granule cells of the dentate gyrus and CA1-CA4 pyramidal cells of the hippocampus, but not in the diagonal band of Broca.[21,22] Lesion studies further confirmed that NGF mRNA was made in neurons.[23] *In situ* hybridization can thus eliminate uncertainties about the cellular sites of altered gene regulation.

Most mRNA species are translated in the cytoplasm surrounding the nucleus, and *in situ* hybridization therefore usually labels cell somata. In most cases, then, *in situ* hybridization bypasses the problem of the identification of the cell with altered gene expression. For example, some receptors, e.g., the $GABA_A$ receptor, are localized both presynaptically and postsynaptically.[24] Alteration in receptor distribution detected with radioactive ligands therefore may represent changes in one or more classes of cells — presynaptic, postsynaptic, or both. *In situ* hybridization with probes for mRNAs for $GABA_A$ receptor polypeptides label

TABLE 1
Alterations in Mutant Mice

Technique	Compound	Area	Cells expressing the gene product	Intrinsic/ extrinsic?	Mutants	Ref.
Ligand binding	Benzodiazepine receptor	Cerebellum	Purkinje cells	Intrinsic	pcd/wv/rl/sg	110
Ligand binding	Histamine H-1 receptor	Cerebellum	Purkinje cells	Intrinsic	pcd/wv/rl/sg	111
Ligand binding	GABA$_A$ receptor	Cerebellum	Granule cells	Extrinsic	pcd/wv/rl/sg	112
Ligand binding	α-1, β-1 adrenoceptors	Cerebellum Hippocampus neocortex	? ? ?	Intrinsic	tg/rl	113,114
Ligand binding	β-1, β-2 adrenoceptors	Cerebellum	?	Intrinsic	rl	115
Immunohistochemistry	Glycerol-3-phosphate dehydrogenase	Cerebellum	Bergmann glia	Extrinsic	Lc-normal/pcd-normal chimeras	116
Immunohistochemistry	L1 cell adhesion molecule	Cerebellum	Granule cells Purkinje cells	Intrinsic	pcd/wv/rl/sg	117,118
Immunohistochemistry	Parvalbumin	Cerebellum	Purkinje, basket, stellate cells	Intrinsic	nr	119
Immunohistochemistry	C1/M1	Cerebellum	Bergmann glia	Extrinsic	pcd/wv/rl/sg	120
Immunohistochemistry	GD3 ganglioside	Cerebellum	Reactive glia	Extrinsic	pcd/wv/Lc/sg	121
Immunohistochemistry	Cerebellin	Cerebellum	Purkinje cells	Extrinsic	pcd/wv/rl/sg/nr	122
Immunohistochemistry	S100	Cerebellum	Bergmann glia	Reduced	rl	123
Immunohistochemistry	Carbonic anhydrase II	Cerebellum	Oligodendrocytes	Increased	rl	123
Immunohistochemistry	UCHT1	Cerebellum	Purkinje cells	Intrinsic	wv/rl/sg	124
Immunohistochemistry	P1-4	Cerebellum	Purkinje cells	Intrinsic	pcd/wv/rl/sg/jp	125
Immunohistochemistry	PEP-19	Cerebellum	Purkinje cells	Intrinsic	pcd/wv/rl/sg/nr	126
Radioimmunoassy	5' nucleotidase	Cerebellum	Bergmann glia	Extrinsic	pcd/hr	127
Histochemistry	Monoamines	Cerebellum	Locus coeruleus, A5, A7 neurons	Intrinsic	pcd	128—130

Method	Gene product	Tissue	Cell type		Mutant	Reference
Biochemistry	pp60 c-src	Cerebellum	?	?	Lc/wv/sg	131
Biochemistry	Dopamine D-2 receptor	Striatum	?	Extrinsic	wv	132
Biochemistry	Dopamine receptor	Cerebellum	?	Extrinsic	rl	133
Biochemistry	Calbindin D_{28K}	Cerebellum	Purkinje cells	Intrinsic	pcd/wv/rl/sg/nr	60
Biochemistry	Nerve growth factor	Cerebellum	?	Extrinsic	pcd	20
Biochemistry	N-CAM	Cerebellum, cortex	Bergmann glia Purkinje cells Granule cells	Extrinsic	wv/rl/sg/jp	18,117
Biochemistry	Particulate neuraminidase	Cerebellum	?	?	wv/sg	134
Biochemistry	Serotonin, tryptophan hydroxylase	Cerebellum	?	Intrinsic	pcd/wv/rl/sg	135
Biochemistry	Thymidine kinase	Cerebellum	Granule cells?	Extrinsic	sg	136
Biochemistry	Galactosyl transferase	Cerebellum	?	Reduced	rl	137
Biochemistry	Hexosaminidase	Cerebellum	?	?	sg/wv	138
Northern blots	MBP, PLP, MAG	Brain	Oligodendrocytes	?	qk/shi/jp	78

cell somata.[25] Thus, *in situ* hybridization can distinguish between changes in pre- and postsynaptic neurons.

3. Quantitation

In situ hybridization can provide direct information about altered gene expression in cells with altered environments. By using radiolabeled probes for specific mRNAs, an investigator can use quantitative autoradiography to estimate the concentration of the corresponding mRNA within individual cells of a particular type. Several groups have already reported quantitative differences in mRNA concentrations.[26,27]

4. Cross-Reactivity

A major problem in immunohistochemistry results from the cross-reactivity of an antiserum with other antigens bearing similar epitopes. Different immunopositive cells thus may not express the same gene product.

A similar problem can occur with *in situ* hybridization. Cross-hybridization due to homologous sequences can occur in multigene families, such as those encoding neurotransmitter receptor polypeptides.

The well-characterized nature of molecular hybridization reactions allows researchers to avoid unwanted cross-hybridization. Increasing the stringency conditions of hybridization or washes may denature mismatched hybrids. In other cases, investigators can avoid cross-hybridization by using oligonucleotide probes directed against unique regions of a gene transcript, particularly in 5′ or 3′ untranslated regions. Once a cDNA is isolated and sequenced, specific oligonucleotide probes are easily synthesized.

The GABA receptor family is an example of a complex class of receptor gene mRNAs exhibiting homology under low stringency conditions, leading to their isolation, but exhibiting different distributions throughout the brain.[28-30]

5. Summary of the Advantages of In Situ Hybridization in the Study of Gene Expression in the Brains of Mutant Mice

In summary, *in situ* hybridization studies provide several advantages for studying specific gene expression in the abnormal brain: (1) since *in situ* hybridization measures changes in mRNA levels, this reveals regulatory changes that occur at the pretranslational level; (2) the specific cells with altered gene transcription in response to the mutated environment can be identified; (3) autoradiography reveals quantitative, as well as qualitative changes in individual mRNAs; (4) sequence-specific probes allow the distinction among cross-reacting sequences, which may otherwise mask changes in gene expression.

D. What the Study of Mutants Can Tell Us About Brain Development

The mouse mutants described in this chapter display developmental abnormalities that affect specific brain regions, cell types, or stages of development. Previous studies, utilizing the techniques discussed above, and the mutants described below, have identified a few gene products whose expression appears to be either intrinsically or extrinsically regulated.

The *staggerer (sg), weaver (wv), Purkinje cell degeneration (pcd), nervous (nr),* and *lurcher (Lc)* alleles result in the degeneration of specific cerebellar neurons. Migration defects leading to cell malpositioning cause the cortical pathology described in the *reeler (rl)* mutants. Genetic defects in oligodendrocytes produce the abnormal myelination observed in the mutants *quaking (qk), jimpy (jp), shiverer (shi),* and its allele, *shi*[mld] (myelin deficient). Finally developmental defects due to chromosomal abnormalities are reported in the mutant, *Trisomy 16 (Ts16).*

II. MUTATIONS AFFECTING CEREBELLAR DEVELOPMENT

The *sg, wv, pcd, Lc, nr,* and *rl* mutations alter the structure of the cerebellar cortex and lead to characteristic ataxias (see Table 2). Here, we briefly describe the development and connections of the normal mouse cerebellum and summarize the cerebellar defects in these six mutants. (For reviews of cerebellar structure and development, see References 31 and 32.)

A. Cerebellar Structure, Connections, and Development
1. Structure
The normal mouse cerebellum contains five major neuronal types, arranged in a trilaminar fashion in the molecular, Purkinje, and granular layers. These neuron types are easily distinguished both by their position and their morphology. Stellate and basket neuron cell bodies occupy, respectively, the outer and inner portions of the cell-sparse molecular layer. Purkinje cell bodies lie in a monolayer, designated the Purkinje layer, between the molecular and granular layers. Numerous small granule neurons and larger Golgi II neurons reside in the granular layer (see Figure 1).

2. Connections
The cerebellar cortex receives two major excitatory inputs, the climbing fibers and the mossy fibers. The climbing fibers arise from neurons of the inferior olivary complex. In the adult cerebellum they innervate Purkinje cells in a one-to-one ratio. During development, however, multiple climbing fibers innervate Purkinje cells. Formation of synapses between Purkinje cells and granule cells reduces this innervation to a single climbing fiber.

Excitatory mossy fibers arise from neurons in the spinal cord, pons, and vestibular nuclei. Mossy fibers synapse with granule cell dendrites at structures called glomeruli. A glomerulus is a bulge in the mossy fiber which is surrounded by granule cell dendrites. Axons of Golgi II neurons impinge upon the glomeruli and provide inhibitory input.

The circuitry within the cerebellar cortex itself is complex. The GABAergic

TABLE 2
Mouse Cerebellar Mutants

Gene	Chromosome	Affected cell types (in cerebellum)	Cells intrinsically affected	Other regions affected
pcd	13	Purkinje cells, inferior olivary neurons	Purkinje cells	Ventromedial thalamus Retinal photoreceptors Olfactory mitral cells
nr	8	Purkinje cells	?	Retinal photoreceptors
sg	9	Purkinje cells, granule cells, inferior olivary neurons	Purkinje cells	None reported
Lc	6	Purkinje cells, granule cells, inferior olivary neurons	Purkinje cells	None reported
wv	16	Purkinje cells, granule cells, Bergmann glia	Granule cells	Striatum
rl	5	Purkinje cells, granule cells	?	Other cortical structures Misplacement of CA3 Pyramidal and dentate granule cells of the hippocampus

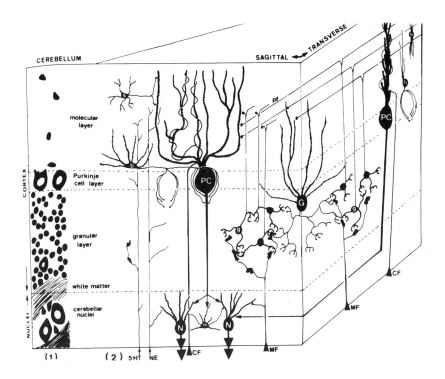

FIGURE 1. Cerebellar structure and connections. (1) The distribution of cells in the molecular, Purkinje layer, granular layer, white matter, and deep cerebellar nuclei are shown as seen in Nissl preparations, which display cell somata. (2) The neurons, their dendrites and axons, of the cerebellar cortex and nuclei as seen in traditional Golgi preparations. b = basket cell, s = stellate cell, PC = Purkinje cell, g = granule cell, G = Golgi II cell, pf = parallel fibers, N = cerebellar nuclear neuron, n = cerebellar interneurons, 5-HT = serotonin afferents, NE = norepinephrine afferents, MF = mossy fiber afferents, CF = climbing fiber afferents. Neurons displayed as they typically appear in the sagittal plane and transverse plane of the brain. (From Chan-Palay, V. and Palay, S. L., in *Encyclopedia of Neuroscience*, Vol. 1, Adelman, G., Ed., Birkhauser Boston, Cambridge, MA, 1987. With permission.)

stellate, basket and Golgi II cells, and the glutamatergic granule cells are interneurons, whose axons terminate only upon other cells within the cerebellar cortex. Stellate cells form inhibitory synapses with Purkinje cell dendrites. Basket cells form synapses with the axon hillock of the Purkinje cell and thus also inhibit Purkinje cell output. The axons of the granule cells (parallel fibers) provide excitatory input to Purkinje, stellate, basket, and Golgi II cell dendrites.

Purkinje cells provide the sole output from the cerebellar cortex, exerting an inhibitory effect through GABA-containing projections to the deep cerebellar nuclei. The deep cerebellar nuclei in turn send projections to the red nucleus. which instructs neurons in the spinal cord, and to the V_A, V_L complex of the thalamus, which regulates neurons in the motor cortex.

3. Development

Many neurons of the cerebellar cortex arise postnatally from the external germinal layer (EGL). Basket neurons undergo their terminal division shortly after the mouse is born. Slightly later, stellate neurons stop dividing. Granule cells arise throughout postnatal development, up until postnatal day 21 (P21). After their terminal division, granule cells send out parallel fibers and their cell bodies migrate along the fibers of the Bergmann glia to reside in the internal granular layer. Golgi II cells, once thought to derive from the ventricular zone on the roof of the fourth ventricle, now appear to be generated from the EGL.[33]

In mice, Purkinje cells and cells of the deep cerebellar nuclei undergo their terminal division between embryonic days 11 to 13 (E11 to 13). They derive from an embryonic ventricular zone on the roof of the fourth ventricle. In rats, the earliest born Purkinje cells derive from the lateral primordium and populate the hemispheres. Purkinje cells born slightly later derive from the postisthmal primordium and populate the posterior vermis. Late-born Purkinje cells derive from the subisthmal primordium and populate the anterior vermis, flocculus, paraflocculus, paramedian lobule of the hemisphere and nodulus.[34] A similar developmental distribution has been reported in mice.[35] The time at which Purkinje cells arise during development thus determines their individual distribution within the cerebellum.

Subsets of developing rat and mouse Purkinje cells can be identified based upon the combinatorial presence or absence of binding of antibodies to calbindin D_{28K}, cyclic GMP dependent protein kinase, and Purkinje cell specific glycoprotein.[36] In the adult, many of these antigens are expressed uniformly in all Purkinje cells. Therefore, "identical" Purkinje cells in the adult derive from subsets of Purkinje cells with different developmental origins and different patterns of gene expression during development. These differences may determine their susceptibility to the different cerebellar mutations.

Indeed, morphological studies of the mouse cerebellar mutants illustrate the different effects of individual mutations upon Purkinje cells. In many cases, there is a gradient of defects (such as cell loss), resulting in more severely affected cells in different regions of the cerebellum. In mutants characterized by Purkinje cell degeneration (*nr, pcd, tambaleante*), more resistant Purkinje cells are arranged symmetrically relative to the midline and in a reproducible pattern specific to each mutation.[37] The *sg* mutation affects Purkinje cells based upon their medial to lateral positions within the cerebellum.[38]

B. Mouse Cerebellar Mutants
1. The Reeler Mutation

The *reeler* mutation produces the earliest reported defects. As early as E14, *rl* embryos have malpositioned neurons.[39] Studies with *rl* <–> wild-type chimeras show that the migration defect is extrinsic to the misplaced neurons.[40,41] Interestingly, proper afferent connections form with many of the malpositioned neurons, showing a separation between the mechanisms governing neuronal position and those governing afferent projections.[42,43]

Purkinje cells lie in several different environments within the *rl* cerebellum. Some Purkinje cells reside in the proper Purkinje cell layer and are bounded by rudimentary molecular and granular layers. These Purkinje cells form synapses with parallel fibers.

Other groups of Purkinje cells lie improperly within the internal granular layer, or remain clustered deep within the cerebellum.[44] In the absence of properly located Purkinje cells, the corresponding granule cell populations are not generated. Thus, these Purkinje cells never form synapses with granule cell axons, and also retain innervation by multiple climbing fibers. The banded pattern of climbing fiber innervation appears normal, however.[42]

The *rl* mutation demonstrates that the juxtaposition of Purkinje cells and the EGL may be necessary for normal mitotic activity in the proliferating EGL. The mitotic activity of the granule cells therefore may be externally regulated.[17] *Reeler* mutants are especially useful to study the effects of cell position upon gene expression.

2. The Staggerer Mutation

Studies of *sg* <–> wild-type chimeras have shown that the mutation acts intrinsically within Purkinje cells.[10] A gradient exists where Purkinje cells in intermediate cerebellar regions are so severely affected by the *sg* mutation that few can be identified. Purkinje cells in medial cerebellar regions are moderately affected by the *sg* mutation.[38]

The *sg* defect results in reduced numbers of Purkinje cells, granule cells, and inferior olivary neurons.[45] The remaining Purkinje cells are stunted, retain multiple climbing fiber innervation, lack tertiary dendritic spines, and do not exhibit calcium spikes.[46] In addition, some Purkinje cells are ectopic.[47]

The lack of tertiary dendritic spines is a primary effect of the *sg* mutation upon Purkinje cells, and is not due to lack of granule cell synapse formation. Purkinje cell spines form independently in the absence of granule cell afferent axons in *rl* and *wv* mutants.[17]

Because *sg* Purkinje cells lack tertiary dendritic spines, they are unable to form synapses with granule cells. The granule cells migrate, then degenerate in the absence of synaptic contact with the Purkinje cells. The granule cell degeneration is a secondary effect resulting from the abnormal *sg* Purkinje cells.[10] Cell loss in the inferior olivary nucleus also appears to be a secondary result of the abnormal Purkinje cells.[45] *Staggerer* mutants can be used to evaluate the importance of synaptic connectivity upon cell survival.

3. The Weaver Mutation

The *wv* mutation is considered semi-dominant because *wv/+* heterozygotes, although behaviorally normal, have abnormal cerebella.[48,49] A defect intrinsic to *wv* granule cells makes them incapable of proper migration.[7-9,50] Many granule cells subsequently die. Cell death is most prominent in medial regions of the cerebellum, where Purkinje cell numbers are also reduced. Some granule cells remain in the hemispheres, however. These granule cells can form proper synapses with Purkinje cells.[49] Bergmann glia demonstrate abnormal processes, but cell culture studies indicate that the abnormal structure depends on interaction with the defective granule cells.[9] Mossy fiber afferents remain, despite the loss of the their targets.[51]

The mechanisms of granule cell migration and the effect upon the cerebellum of the loss of the excitatory granule cells can be evaluated in *wv* mutants.

4. The Purkinje Cell Degeneration Mutants: Lurcher, Nervous, and Purkinje Cell Degeneration

The primary lesion occurs in the Purkinje cells of *pcd* and *Lc* mutants, resulting in Purkinje cell degeneration at later stages of development.[12,52] Studies of *nr* <–> wild-type chimeras were inconclusive regarding the primary cellular lesion.[53]

The Purkinje cells of *pcd* mutants appear abnormal starting at P15. Purkinje cell loss begins around P17 and progresses rapidly until less than 1% of the Purkinje cells remain by P28. Remaining Purkinje cells are found in the nodulus, flocculus, paraflocculus, and uvula vermis. The numbers of neurons in the inferior olive are also less than in wild-type mice.[54]

Prior to their degeneration, *Lc* Purkinje cells have been reported to have more than one primary dendrite (sometimes up to five) and somatic spines that persist abnormally beyond the first postnatal week. Other abnormalities include rounded mitochondria (like those seen in *nr* Purkinje cells) and randomly oriented organelles. Purkinje cell loss begins on P8, and by P26 fewer than 10% of the Purkinje cells remain. *Lurcher* mutants also lose granule cells and inferior olivary neurons.[55]

All Purkinje cells in P15 *nr* cerebella contain abnormally rounded mitochondria. The mitochondria in the surviving 10% of the Purkinje cells demonstrate their normal elongated shape after having passed through intermediate stages. The remaining Purkinje cells are in the vermis.[56]

C. *In Situ* Hybridization Studies in Mouse Cerebellar Mutants

In situ hybridization studies have revealed the effects of altered connections and cellular environment upon gene expression in the cerebellum. The mutants examined include *rl, Lc, sg,* and *wv*.

1. Glutamate Decarboxylase, Proenkephalin, and Calbindin D_{28K} mRNAs in Weaver and Reeler Cerebella

Wuenschell and Tobin examined the expression of three genes in *rl* and *wv* cerebella: calbindin D_{28K}, glutamate decarboxylase (GAD), and proenkephalin. Since these genes encode, respectively, a calcium binding protein likely to function postsynaptically, a neurotransmitter synthetic enzyme, and a neuropeptide precursor, they are likely to play an essential role in neuronal function. Only Purkinje cells contain calbindin D_{28K} mRNAs. GAD mRNAs exist in stellate, basket, Golgi II, and Purkinje cells.[57] In the cerebellar cortex, Golgi II cells and possibly basket cells contain proenkephalin mRNAs. Thus, the two largest neuron types in the cerebellar cortex can be distinguished by their pattern of gene expression: Golgi II cells contain GAD and proenkephalin mRNAs, and Purkinje cells contain GAD and calbindin D_{28K} mRNAs.

Despite the absence of granule cells in the *wv* cerebellum, the expression of calbindin D_{28K} and GAD mRNAs in the Purkinje cells, and of proenkephalin and GAD mRNAs in the Golgi II cells is unaffected. In the *rl* cerebellum, GAD and calbindin D_{28K} mRNAs are detected in the Purkinje cells located in the rudimentary cerebellar cortex (forming synapses with granule cells) and clustered in the deep regions (lacking synapses with granule cells). The authors concluded that Purkinje cells express these two genes autonomously, even in the absence of afferent input from the granule cells[58] (see Figure 2).

2. Glutamate Decarboxylase and Calbindin D_{28K} mRNAs in the Lurcher Cerebellum

In a second study, Wuenschell et al. examined GAD and calbindin D_{28K} mRNAs in the cerebella of *Lc* mice, in which Purkinje cells first develop normally, in an apparently normal environment, and then die. All GABAergic neuron types in the P21 *Lc* cerebella have normal levels of GAD mRNAs and appear to be unaffected by the already evident Purkinje cell loss. The Purkinje cells still remaining at P21 contain both GAD and calbindin D_{28K} mRNAs. Aberrant expression of these two genes thus does not contribute to Purkinje cell degeneration. The *Lc* mutation also appears to have few indirect effects upon the expression of other genes in Purkinje cells. Therefore, Purkinje cell death in *Lc* mice apparently results from the isolated action of the mutant gene, rather than from a cascade of aberrant gene expression triggered by the mutation[59] (see Figure 3).

3. Glutamate Decarboxylase and Calbindin D_{28K} mRNAs in Staggerer Cerebella

Like *Lc*, the *sg* mutation acts in Purkinje cells themselves, affecting their final forms and connections. The expression of several Purkinje cell specific genes, among these calbindin D_{28K}, may be reduced in the remaining *sg* Purkinje cells. Parkes et al.[60] detected only very low levels of calbindin D_{28K} in radioimmunoassays performed on *sg* cerebella. Likewise, Heinlein et al.[61] report diminished levels of a 28 to 29 kDa protein in protein gels containing *sg* cerebella samples, which they presume is calbindin D_{28K}. Frantz et al. used *in situ* hybridization to study GAD and calbindin D_{28K} mRNAs in *sq* mice.

The Purkinje cells of *sg* mice contain both GAD and calbindin D_{28K} mRNAs. The ectopic Purkinje cells of *sg*, however, express greatly reduced levels of calbindin D_{28K} mRNAs compared to orthotopic Purkinje cells. Since granule cells do not make synaptic contact with any *sg* Purkinje cells, the loss of granule cells does not appear to cause this difference.

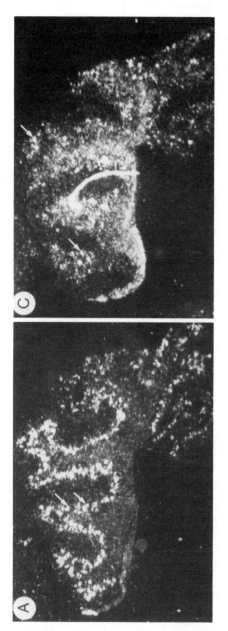

FIGURE 2.   Calbindin D$_{28K}$, GAD, and proenkephalin mRNA distribution in the cerebella of *reeler* and *weaver* mice. (I) Darkfield photomicrographs of sections of *weaver* cerebellum hybridized with antisense GAD RNA (A), antisense calbindin D$_{28K}$ RNA (B), antisense proenkephalin RNA (C), or sense calbindin D$_{28K}$ RNA (control, D). A, B, and C: Arrows indicate labeled objects in the cortical layer. D: Arrows indicate locations equivalent to those marked in (A), (B), and (C). (II) Darkfield photomicrographs of sections of *reeler* cerebellum hybridized with antisense GAD RNA (A), antisense calbindin D$_{28K}$ RNA (B), antisense proenkephalin RNA (C), or sense calbindin D$_{28K}$ RNA (control, D). A: Arrows point to labeled objects at the boundary between the molecular and granular layers. B: Arrow at left points to a labeled object in the granular layer. Arrows at right point to labeled objects at the boundary between the molecular and granular layers. C: Arrows at left point to labeled objects in the deep nucleus. Arrow at right points to labeled objects in the granular layer. c = central cell mass, d = deep nucleus, g = granular layer.

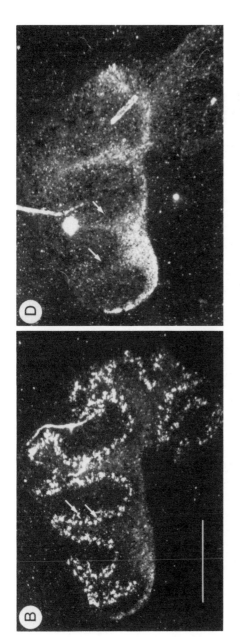

FIGURE 2.

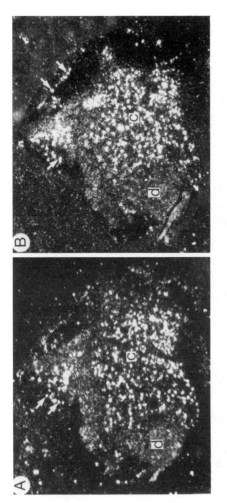

FIGURE 2 (continued).

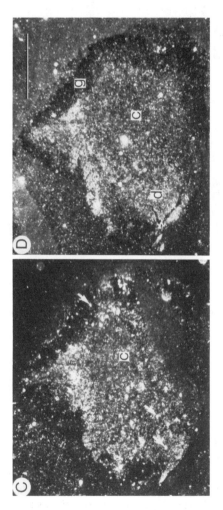

FIGURE 2 (continued).

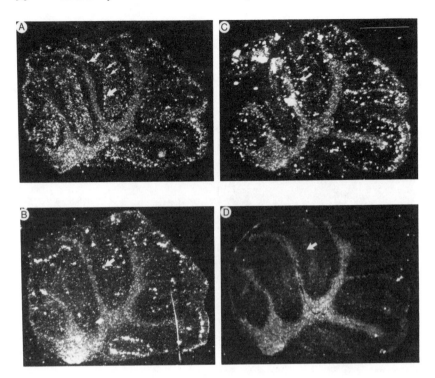

FIGURE 3. Calbindin. A: Large arrow shows labeled Purkinje cell. Curved arrow shows labeled Golgi II cell. Small arrows show labeled basket and stellate cells. B: Large arrow shows labeled Purkinje cell. C: Curved arrow shows labeled Golgi II cell. Small arrows show labeling in the deep molecular layer. D: Large arrow delineates the Purkinje layer.

Ectopic Purkinje cells may not make other appropriate connections, however. On the other hand, ectopic Purkinje cells may be more severely affected by the *sg* mutation, as shown by their abnormal position. In these cells calbindin D_{28K} mRNA levels may be reduced as an indirect effect of the *sg* mutation[62] (see Figure 4).

4. N-CAM mRNAs in the Cerebella of Reeler and Lurcher Mice

Almost all of the neurons in the cerebellum and hippocampus of normal mice contain high levels of mRNA for the neural cell adhesion molecule, N-CAM. Two cell types, however, lack detectable N-CAM mRNA: (1) cerebellar granule cells during their migration from the EGL, and (2) granule neurons of the dentate gyrus just after their final round of DNA synthesis. These exceptions suggest that N-CAM plays a greater role in later developmental events, such as axon fasciculation and synaptogenesis, than in early developmental events, such as cell migration.

In situ hybridization showed that N-CAM mRNA levels in cerebellar cells do not differ between normal and *rl* mice. Purkinje cell location thus does not affect N-CAM expression in Purkinje cells themselves, or in other cerebellar neurons.

The granule cells of the *Lc* cerebellum have decreased levels of N-CAM mRNAs, however. One explanation could be that granule cells require continued contact with their efferent targets (the Purkinje cells) to maintain constant N-CAM expression. N-CAM mRNA levels in the surviving *Lc* Purkinje cells are unaffected.[63]

5. Calmodulin mRNAs in Staggerer and Lurcher Cerebella

Many neuronal functions are regulated through the binding of calcium to the intracellular protein, calmodulin.[64] Both immunohistochemical and *in situ* hybridization studies demonstrate high levels of calmodulin and calmodulin mRNAs in the Purkinje cells of the cerebellar cortex.[65-67]

Messer et al. studied the distribution of calmodulin mRNAs in developing normal and *sg* cerebella. In wild-type mice, calmodulin mRNA localization varies during development. In newborn mice, calmodulin mRNA appears in both the molecular and Purkinje layers. By P14, Purkinje cells have much higher levels of calmodulin mRNA than the cells of the molecular layer. The labeled cells in the molecular layer of the wild-type cerebellum are migrating granule cells. In adult mice, the Purkinje cells continue to show labeling for calmodulin mRNA, but the molecular layer does not. Instead, diffuse labeling appears in the internal granular layer where the migrating granule cells have settled.[68]

The cerebella of *sg* mice show no labeling of Purkinje cells with calmodulin probes. *Staggerer* granule cells, however, express near normal levels of calmodulin mRNA, up until the time they degenerate. This indicates that the defect in calmodulin expression in the *sg* cerebellum is specific to the Purkinje cells.

In contrast, the remaining Purkinje cells of P18 *Lc* mice appear to express normal amounts of calmodulin mRNA.[69]

D. Summary of *In Situ* Hybridization Studies in Mouse Cerebellar Mutants

The above studies reveal several intriguing features of gene expression in Purkinje cells. The expression of the GAD, calbindin D_{28K}, and N-CAM genes in *rl* and *wv* Purkinje cells does not appear to depend upon proper position or input from granule cells. This suggests intrinsic regulation of the expression of these genes in Purkinje cells.

Lurcher Purkinje cells express GAD, calbindin D_{28K}, N-CAM, and calmodulin mRNAs normally, until they die. Thus, the degeneration of *Lc* Purkinje cells does not result from an overall flaw in gene expression.

In contrast, the *sg* mutation appears to disrupt gene expression in Purkinje cells. Calbindin D_{28K} mRNA levels in some *sg* Purkinje cells are greatly reduced, while calmodulin mRNA levels are below the level of detection in *sg* Purkinje cells, but not in *sg* granule cells (see Table 3).

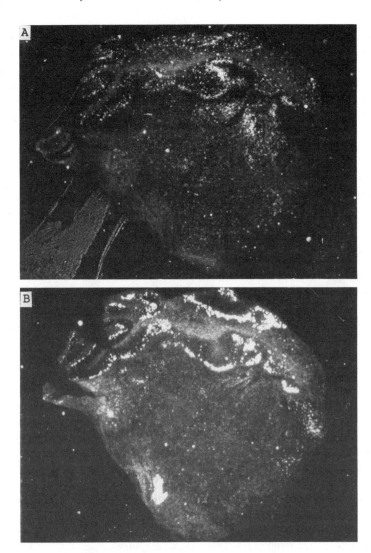

FIGURE 4.   Calbindin D and GAD mRNA distribution in the cerebella of *Staggerer* mice. Darkfield photomicrographs of sections of P21 *staggerer* cerebellum hybridized with antisense GAD RNA (A) or antisense calbindin

III. MUTATIONS AFFECTING HIPPOCAMPAL DEVELOPMENT

A. Hippocampal Structure and Development

The structure of the hippocampus, like that of the cerebellum, is laminar, with distinct cell types distributed within defined layers. In Ammon's horn these layers are the oriens, pyramidal, radiatum, and lacunosum moleculare. The 3 to

TABLE 3
In Situ Hybridization Studies in Mouse Mutants

Mutant	Cell type/area examined	Gene(s) examined	Level of Expression (compared to control)	Extrinsic/ intrinsic?	Ref.
wv	Golgi II/cerebellum	GAD, proenkephalin	Unaffected	Intrinsic	58
	Purkinje/cerebellum	GAD, calbindin D	Unaffected	Intrinsic	58
rl	Purkinje/cerebellum	GAD, calbindin D, N-CAM	Unaffected	Intrinsic	58,63
	Golgi II/cerebellum	GAD, proenkephalin	Unaffected	Intrinsic	58
	CA3 pyramidal/hippocampus	SNAP-25	Unaffected	Intrinsic	75
sg	Purkinje/cerebellum	GAD	Unaffected	Intrinsic	62
	Purkinje/cerebellum	Calbindin D, calmodulin	Decreased	?	62,69
	Granule/cerebellum	Calmodulin	Unaffected	Intrinsic	69
Lc	Purkinje/cerebellum	GAD, calbindin D, calmodulin, N-CAM	Unaffected	Intrinsic	59,63. 69
	Granule/cerebellum	N-CAM	Decreased	Extrinsic	63
qk	Oligodendrocyte	MBP	Unaffected	Intrinsic	94
JCV	Oligodendrocyte	MBP, PLP	Unaffected	Intrinsic	93
	Astrocyte	GFAP	Increased	Extrinsic	93
Ts16	Fetal brain	SMST, GAP-43, Ets-2, APP	Increased	?	97
Ts19	Fetal brain	GAP-43	Decreased	?	97
	Fetal brain	APP	Unaffected	?	97

5 cell thick pyramidal layer of Ammon's horn contains the cell bodies of the pyramidal cells, the principal neurons of hippocampal fields CA1 to CA3. The pyramidal cells in field CA1 are smaller and more tightly packed than those in fields CA2 and CA3. The oriens layer of Ammon's horn contains the basal dendrites of the pyramidal cells. The apical dendrites of the pyramidal cells extend through the radiatum layer to the lacunosum moleculare layer. The pyramidal cells are generated prenatally and migrate to their final positions.

The dentate gyrus consists of the molecular and granular layers and the hilus. The granular layer contains granule cell somata which are densely packed 4 to 10 cells thick. Granule cell dendrites extend into the molecular layer. Granule cells are generated postnatally.

The hippocampus connects directly or indirectly with many brain regions. The major input to the hippocampus comes via the perforant path, which arises in the entorhinal cortex and terminates in the lacunosum moleculare layer of field CA3 and the outer $^2/_3$ of the molecular layer of the dentate gyrus. The axons of dentate granule cells (mossy fibers) project through the hilus and terminate on CA3 pyramidal neurons. The axons of both CA1 and CA3 pyramidal cells project via the fornix to the septum. These connections form the "trisynaptic pathway" of the hippocampus and dentate gyrus (see Figure 5). For a review of hippocampal structure, see Reference 70.

B. The *Reeler* Mutation

The *rl* mutation affects the organization of the hippocampus, as well as of the cerebellum. The *rl* hippocampus contains cells that have failed to migrate to their proper positions. In *rl* some pyramidal cell bodies lie normally in the pyramidal layer, but others are dispersed above this layer. In addition, many of the *rl* pyramidal cells are inverted. Because these cells are "upside down", inputs which normally terminate, for example, on outer dendritic segments, now terminate on inner dendritic segments, and vice-versa.

Similarly, the granule cells of the *rl* dentate gyrus are not tightly packed into the granular layer, but are more dispersed, with some granule neurons lying within the hilus.[71] Many of these ectopic cells nonetheless make proper synaptic connections.[72,73] As previously described in studies of the cerebellum, the *rl* mutation provides an opportunity to evaluate the influence on gene expression of cell migration and connections.

C. SNAP-25 Expression in *Reeler* Mice

A 25-kDa protein called SNAP-25 (*syn*aptosomal *a*ssociated *p*rotein), is associated with specific synapses in the hippocampal pathways described above. SNAP-25 is in presynaptic nerve terminals in the perforant path, mossy fiber projections, and projections to the septal nucleus.[74] In normal mice, *in situ* hybridization reveals high levels of SNAP-25 mRNA in the cell bodies of CA3 pyramidal cells.

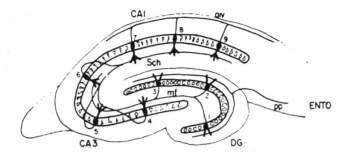

FIGURE 5. Hippocampal structure and connections. A schematic diagram of the principal trisynaptic circuit of the hippocampus. ENTO = entorhinal cortex, pp = perforant path, DG = dentate gyrus, mf = mossy fibers, Sch = Schaffer collaterals, alv = alveus. (From Teyler, T. J. et al., in *The Hippocampus*, Vol. 3, Isaacson, R. L. and Pribram, K. H., Eds., Plenum Press, New York, 1986. With permission.)

Wilson et al. have also examined the distribution of SNAP-25 mRNA in the *rl* hippocampus. In *rl*, some of the CA3 pyramidal cells have not migrated to their final positions, but nonetheless contain high levels of SNAP-25 mRNA. In addition, immunohistochemical studies reveal SNAP-25 within the mossy fiber projections, which terminate on the apical dendrites of CA3 pyramidal cells (see Figure 6). These results demonstrate that SNAP-25 expression is regulated independently of cell position. The induction of SNAP-25 expression is thus likely to depend on other developmental interactions, such as proper innervation or trophic factors that induce neurite extension.[75]

IV. MYELINATION MUTATIONS

A. Gene Expression during Myelination

Myelination in the central nervous system (CNS) begins postnatally with the compaction of many layers of oligodendrocyte cell membranes around axons. Oligodendrocytes synthesize proteolipid protein (PLP), myelin basic protein (MBP), and myelin associated glycoprotein (MAG), which are important components in this process. In adult animals these three proteins represent, respectively 50%, 30 to 40%, and 1% of total myelin protein. Isoforms of MBP, PLP, and MAG result from alternative splicing of RNA transcripts.

MAG appears to function in the adhesion of oligodendrocytes to axons prior to myelination. MBP first appears during myelin compaction, when it enters the cytoplasmic face of the membrane. Interactions between MBP molecules appear to keep in close apposition the cytoplasmic faces of myelin. PLP molecules, on the other hand, appear to mediate the apposition of the extracellular membranes. For a review of myelin genes, see References 76 and 77.

The developmental regulation of the PLP and MBP genes is complex. In

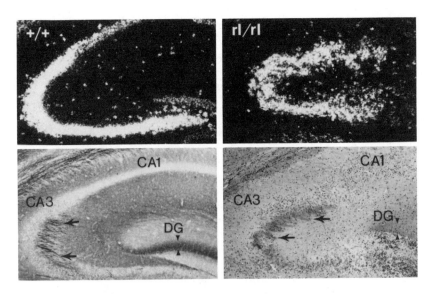

FIGURE 6.   SNAP-25 mRNA distribution in the hippocampus of *reeler* mice. *In situ* hybridization and immunohistochemical localization of SNAP-25 gene expression in the hippocampal formation of normal and *reeler* mice. Upper panels are darkfield photomicrographs of sections of normal and *reeler* hippocampi hybridized with antisense SNAP-25 RNA. Lower panels are alternate sections stained with antibody raised against a synthetic peptide encoded by the carboxy-terminus of the SNAP-25 sequence. (Hart, R. A. and Wilson, M. C., unpublished observations.)

normal mice, PLP and MBP mRNAs rapidly accumulate in the nuclei of oligodendrocytes, until they reach their maximal levels at P18. Then levels of nuclear PLP mRNAs drop, but levels of cytoplasmic PLP mRNA remain high. Nuclear levels of MBP mRNA fall, then rise again, but cytoplasmic MBP mRNA levels fall.[78] The expression of the MBP gene appears to be intrinsically regulated: cultures of oligodendrocytes constitutively make myelin, regardless of the presence or absence of neurons.[79]

In situ hybridization studies in normal mice reveal that PLP and MBP mRNAs have distinct cellular distributions. PLP mRNAs are associated with ribosomes bound to the endoplasmic reticulum within cell bodies. In contrast, MBP mRNAs are associated with free ribosomes that move from the cell somata to the oligodendrocyte processes early in myelogenesis.[80,81]

B. Mice with Defects in Myelination

Mice with myelination defects have tremors, tonic convulsions, and ataxia. The mutants discussed here include *shi, shi*[mld], *jp, qk,* and dysmyelinating transgenic mice (see Table 4).

1. The Shiverer and Shiverer[mld] *Mutations*

The autosomal recessive mutations, *shi* and *shi*[mld], map to chromosome 18.

TABLE 4
Mutants with Myelination Defects

Mutant	Chromosome	Gene affected
shi	18	MBP
shi[mld]	18	MBP
jp	X	PLP
qk	17	?
JCV	Introduced	T-antigen

Shiverer mutants exhibit tremors and die by age 90 to 150 d. The *shiverer* allele results from a deletion of exons 3 through 7 of the MBP gene.[82] Interestingly, PLP mRNA levels in these mutants are also reduced to 30 to 55% of control levels.[78]

The *shi*[mld] mutation results from a duplication of the MBP gene that contains an inversion.[83] This causes a depression of MBP expression, but the MBP mRNA is of normal length. The *shi*[mld] allele produces a less severe mutation than the *shi* allele. By P90, MBP levels in *shi*[mld] reach 10% of control levels.[84] In this case, *in situ* hybridization studies might reveal the effects of MBP mutations upon the intracellular partitioning of the MBP mRNAs.

2. The Jimpy Mutation

The *jp* mutation is X-linked recessive. The *jp* mutant displays tremors at P11 and dies by P25 to P30. A point mutation in the splice acceptor sequence of exon 5 of the mouse PLP gene results in a truncated PLP protein.[85,86]

The *jp* mutation also causes a reduction in the number of oligodendrocytes and MBP and MAG content.[87,88] MAG mRNA levels are reduced to less than 10% of control levels.[89] One group has reported a reduction in total MBP mRNAs, while another has reported differences in the relative proportions of mRNAs for the high and low molecular weight MBP isoforms.[78,90]

The decreased levels of MBP and MAG mRNAs may result from reduced expression of these specific genes within the remaining oligodendrocytes. Alternatively, the expression of these genes may remain normal in the oligodendrocytes, and the decreased mRNA levels may simply reflect the loss of oligodendrocytes. *In situ* hybridization could determine whether the reduced levels of MBP and MAG RNAs are the result of decreased numbers of oligodendrocytes, or a specific reduction in the RNAs of these specific genes.[76]

3. The Quaking Mutation

The *qk* autosomal recessive mutation maps to chromosome 17. *Quaking* mutants exhibit tremors beginning at P10, but have a relatively normal life span. MBP, PLP, and MAG are all affected by the *qk* locus and the brain of a *qk* homozygote contains only 5 to 10% of the wild-type levels of myelin.[91] The most anterior regions of the *qk* brain exhibit the most severe dysmyelination.

Quaking mutants have alterations in the glycosylation of MAG. In addition, *qk* mice have altered expression of the MAG gene. The predominant form of

MAG mRNA in young (P15 and P25) *qk* mice resembles the mRNA present only in older wild-type mice.[89] PLP mRNA levels are also reduced in *qk* brains.[78] Neonatal animals exhibit lower MBP mRNA levels than controls.[92] By P21 MBP mRNA levels are normal, but steady state levels of the proteins themselves are only 5 to 25% of wild-type, suggesting defective incorporation of MBPs into myelin.[78]

4. Dysmyelinating Transgenic Mice

The papovavirus, JC virus (JCV), causes the demyelinating disease, progressive multifocal leukoencephalopathy in humans. To study this disease, Trapp et al. created JCV transgenic mice as a model system. These mice contained the JCV early region, which encodes the large and small T-antigens and contains the cell-specific JCV enhancer/promoter region. The mice did not receive the viral late sequences, necessary for productive viral infection.[15]

These transgenic mice demonstrated neurological disorders resulting from CNS dysmyelination. The transgenic mice exhibited dysmyelination (defective formation of myelin) rather than demyelination (degeneration of myelin). The severity of dysmyelination is correlated with the level of JCV T-antigen mRNA in oligodendrocytes.[93]

C. *In Situ* Hybridization in Mice with Myelin Defects
1. MBP Expression in Quaking Mice

Since MBP maps to chromosome 18 and the *qk* allele to chromosome 17, the *qk* defect lies outside of the MBP gene itself. The inability to incorporate MBPs into *qk* myelin, however, could be due to a defect in the translocation of MBP mRNAs within the oligodendrocyte, or to some other post-translational event. *In situ* hybridization reveals no differences in the levels or intracellular distribution of MBP mRNAs between *qk* and normal mice. These results indicate a defect in the assembly of MBPs into the membrane, subsequent to the movement of the MBP-synthesizing ribosomes from the oligodendrocyte somata to the processes.[94]

2. MBP, PLP, GFAP, and T-Antigen Expression in Dysmyelinating Transgenic Mice

In situ hybridization in JCV transgenic mice shows normal PLP and MBP mRNA levels in oligodendrocytes. In contrast, PLP and MBP immunoreactivities are greatly reduced compared to control animals. These results suggest (1) that dysmyelination in the transgenic mice depends upon the expression of JCV T-antigens in oligodendrocytes, and (2) that the maturation of oligodendrocytes arrests at an early stage. Myelin genes produce mRNA, but myelin components do not properly assemble.

Astrocytes in the JCV transgenic mice are also affected, as well as oligodendrocytes. Antibodies against glial fibrillary acidic protein (GFAP) stained astrocytes, but these astrocytes were larger than normal. They also contained far more GFAP mRNA than normal astrocytes. Such astrocyte hypertrophy appears to be a general response to CNS hypomyelination. Thus, the expression of JCV T-antigen appears to affect the cytodifferentiation of oligodendrocytes directly and that of astrocytes only indirectly.[93]

V. TRISOMY 16

A. Description of Ts16

Trisomy 16 (Ts16) is a lethal condition, in which affected mice have three copies of chromosome 16. Many of the genes on mouse chromosome 16 are synthetic with genes located on human chromosome 21. In humans, trisomy 21 (Down syndrome) leads to characteristic mental retardation and abnormal development. Down syndrome patients have decreased brain size, fewer cortical convolutions, lower numbers of cortical neurons, and reduced neuronal dendritic complexity. Down syndrome patients almost invariably develop the neuropathology characteristic of Alzheimer's disease. *Trisomy 16* in mice has therefore generated considerable interest as a possible model for Down syndrome and Alzheimer's disease.[95]

Trisomy 16 mice have specific reductions in the cholinergic, noradrenergic, serotonergic, and dopaminergic, but not the GABAergic neurotransmitter systems. Aneuploidy usually disrupts embryogenesis, and these changes may be the nonspecific result of abnormal chromosome number. On the other hand, these changes may result from the dosage effects of specific genes on chromosome 16. Comparisons between *Ts16* and *Ts19* fetuses demonstrate that reduced brain weight and protein content is a general consequence of aneuploidy. The neurochemical content of *Ts19* brains, however, differs from that of *Ts16* brains, indicating that the reductions in neurotransmitter systems are unique to *Ts16*.[96]

B. *In Situ* Hybridization in *Trisomy 16* Mice

Chromosome 16 carries the genes for preprosomatostatin (SMST), the 43 kDa growth associated phosphoprotein (GAP-43), amyloid precursor protein (APP), and the oncogene Ets-2. These genes encode, respectively, a neuropeptide precursor, a phosphoprotein associated with axonal elongation and synaptic modification, an extracellular component of the senile neuritic plaques of Alzheimer's disease, and a protooncogene. Both Ets-2 and APP map to human chromosome 21, while SMSt and GAP-43 map to human chromosome 3. In normal animals, *in situ* hybridization experiments and RNA blots demonstrate all four mRNAs between E10 to E12.[97]

In situ hybridization and RNA blot experiments show that the gene dosage at least partly determines gene expression. *Ts16* mice fetuses have increased levels of mRNA for SMSt, Ets-2, APP, and GAP-43.[81] For example, at E15 GAP-43 mRNA levels were 2 to 3-fold higher than normal in *Ts16* brains, but only 80% of normal in *Ts19* brains.[98]

Similarly, embryonic *Ts16* mice demonstrate up to a threefold increase in

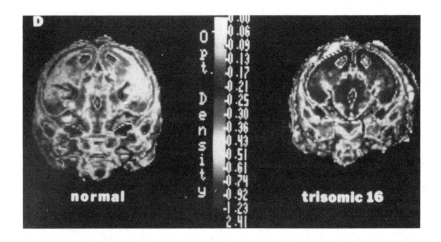

FIGURE 7.   APP mRNA distribution in the brains of fetal normal and *Ts16* mice. Computer images of the expression of APP mRNA in the heads of normal (left) and *Ts16* (right) mice at E15. Note that the OD varies from 0.00 to 2.41. There is a clear increase in hybridization signal, but a similar distribution of expression in various brain regions of *Ts16*, relative to a control littermate fetal mouse brain.

APP mRNA in specific regions, e.g., the cerebral cortex. *Ts19* fetuses, however, showed a similar APP mRNA distribution to wild-type fetuses throughout the developing brain (see Figure 7). RNA blot analysis of the entire *Ts16* mouse brain reveals a threefold increase in APP mRNA, far greater than that expected from the 1.5-fold gene dosage effect.[95,99]

Transcriptional levels of any gene in aneuploid mice, thus cannot be predicted entirely by gene copy number. The features unique to *Ts16* mice, however, probably result from altered gene transcription in response to altered gene copy number.

VI. FUTURE PROSPECTS

A. Neuron-Specific Genes

Mouse neurological mutants also provide a means of identifying genes expressed in specific neuron types. Two groups, for example, have used the cerebellar mutants *Lc* and *pcd* to identify cDNA probes for mRNAs specifically expressed in Purkinje cells.

1. Purkinje Cell Specific Genes

Oberdick et al. identified two Purkinje cell specific cDNAs, L7 and L19.[100] They screened a P13 mouse cerebellar cDNA by differential colony hybridization, using cDNA probes copied either from the mRNAs of wild-type (Purkinje cell +) or of *Lc* (Purkinje cell –) cerebella. The sequence of L19 corresponds to that of PEP-19,[101] while L7 appears to represent a novel Purkinje cell specific mRNA. Both L7 and L19 mRNAs appear postnatally, and *in situ* hybridization demonstrates their presence within Purkinje cells.[100] Surprisingly, L7 RNA is present not only in somata, but also in dendrites.[102]

Nordquist et al. also identified Purkinje cell specific mRNAs by using "subtracted" cDNA probes. These cDNAs contain sequences that are present in normal, but not in *pcd* mutant mice. They identified three Purkinje cell specific cDNAs, which they called PCD5, PCD6, and PCD29. The sequences of PCD5 and PCD6 did not match those of any reported proteins, while that of PCD29 matched that of calbindin D_{28K}. *In situ* hybridization showed the presence of the corresponding mRNAs in Purkinje cells.[103]

2. The rds Gene

The isolation of cDNAs for mRNAs expressed in specific neuron types can lead to the identification of gene mutations producing certain abnormal phenotypes. For example, Travis et al. have used this approach to identify the gene responsible for the *retinal degeneration slow* (rds) phenotype. The expression of the *rds* locus within retinal photoreceptor cells leads to their degeneration. Therefore, the *rds* gene was predicted to encode an mRNA expressed only in photoreceptor cells.

The *retinal degeneration (rd)* mutant, which like *rds* lacks photoreceptor cells, should lack mRNAs that are confined to photoreceptors. "Subtraction" of a mixture of *rd* retinal cDNAs from wild-type retinal cDNAs thus gives a mixture of cDNAs enriched in photoreceptor specific sequences. Use of such a subtracted cDNA mixture allowed the isolation of 12 cDNA clones representing photoreceptor mRNAs. One of these, called B9A, hybridized to an RNA whose molecular weight was much larger in *rds* retinas (before the onset of photoreceptor degeneration) than in wild-type retinas. B9A and *rds* both map to chromosome 17. Genomic clones isolated from *rds* mice demonstrate an inserted element within the coding sequence of the B9A gene. The normal *rds* gene encodes a peptide of 346 amino acids.[104]

B. Genes Regulating Development

The products of many genes, including protooncogenes, growth factors, and homeobox genes, contribute to the regulation of development. Altered expression of these genes may affect the expression of specific sets of genes. cDNAs have been isolated for many of these genes, and *in situ* hybridization studies in normal mice have demonstrated that their expression varies during neural development.

Studies of the expression of these genes in normal mice suggest that protooncogene, growth factor, and homeobox gene expression may be important determinants of neural development. At this printing, however, no data are available on alterations in expression in neurological mutants.

1. Protooncogene Expression during Cerebellar Development

Ruppert et al. studied the developmental expression of the protooncogene c-myc in the mouse cerebellum. The c-myc product apparently functions through its DNA-binding ability. Prior to birth, the cerebellum contains high levels of c-myc transcripts. After birth, these levels fall, then rise again. c-myc appears postnatally in proliferating cells of the EGL of P3 mouse cerebella. Later, c-myc mRNA appears both in proliferating EGL cells and in differentiating Purkinje cells of P10 mouse cerebella. Some adult Purkinje cells also contain low levels of c-myc mRNA, but none is detected in the mature granule cells. These results suggest a role for c-myc in proliferative and differentiation events during cerebellar development.[105]

In contrast, mRNAs for the mammary tumor protooncogene Int-1 appear prenatally, in the neural plate cells of E9 mice. Before E14, Int-1 mRNA appears in several brain regions. Int-1 may thus play a role in the morphogenesis of the neural tube or early brain development.[106]

2. Growth Factor mRNAs in the Brain

Transforming growth factor alpha (TGFα) has structural homology with epidermal growth factor (EGF) and binds to the EGF receptor. TGFα plays a role during early fetal development. *In situ* hybridization reveals TGFα mRNA in many regions of the adult mouse brain. These same brain regions make EGF and enkephalins.[107]

3. Homeobox Gene Expression during Neural Development

En-2 is a homeobox containing gene. En-2 mRNAs are first detected using *in situ* hybridization in the neural folds of E8 mice. At E8.5, En-2 mRNAs exist in all cells of the future midbrain and hindbrain, but by E12.5, En-2 expression is restricted to cells of the metencephalon. In adult mice, cerebellar granule cells and some cells in the molecular layer contain En-2 transcripts.[108]

The homeobox genes, Hox 2.1 and 2.2, are also expressed in developing neural structures. Hox 2.1 and 2.2 mRNAs exist in E13.5 mouse embryos in cells of the developing hindbrain and spinal cord. Hox 2.1 expressing cells extend more rostral than Hox 2.2 expressing cells.[105]

VII. SUMMARY

Many inherited neurological disorders in humans may result from abnormal gene expression in the brain. Because of the complex relationships among brain cells, abnormal gene expression in one cell type can have dramatic effects upon other cell types. *In situ* hybridization studies of mouse neurological mutants and transgenic mice can lead to a better understanding of normal gene expression and its perturbation in neurological and neuropsychiatric disorders.

ACKNOWLEDGMENTS

We thank Dr. Anthony Campagnoni, Dr. Joseph Coyle, Dr. Daniel Goldowitz, Dr. Anne Messer, Dr. Harold Orr, Neil Verity, Dr. Michael Wilson, and Dr. Carol Wuenschell for sharing published and unpublished observations; Dr. Timothy Teyler and Dr. Victoria Chan-Palay for permission to reproduce artwork; and Herman Kabe for photographic work. This work was supported by a grant to A. J. T. from NINCDS (#NS 20356). G. D. F. was supported in part by an NIH training grant in cellular and molecular biology (#GM 07185).

REFERENCES

1. **Altman, J. and Anderson, W. J.,** Experimental reorganization of the cerebellar cortex. I. Morphological effects of elimination of all microneurons with prolonged X-irradiation started at birth, *J. Comp. Neurol.,* 146, 355, 1972.
2. **Sotelo, C. and Alvarado-Mallart, R. M.,** Embryonic and adult neurons interact to allow Purkinje cell replacement in mutant cerebellum, *Nature,* 327, 421, 1987.
3. **Takayama, H., Kohsaka, S., Shinozaki, T., Inoue, H., Toya, S., Ueda, T., and Tsukada, Y.,** Immunohistochemical studies on synapse formation by embryonic cerebellar tissue transplanted into the cerebellum of the *weaver* mutant mouse, *Neurosci. Lett.,* 79, 246, 1987.
4. **Buzsaki, G., Ponomareff, G., Bayardo, F., Shaw, T., and Gage, F. H.,** Suppression and induction of epileptic activity by neuronal grafts, *Proc. Natl. Acad. Sci. U.S.A.,* 85, 9327, 1988.
5. **Savage, D. D., Rigsbee, L. C., and McNamara, J. O.,** Knife cuts of entorhinal cortex: effects on development of amygdaloid kindling and seizure-induced decrease of muscarinic cholinergic receptors, *J. Neurosci.,* 5, 408, 1985.
6. **Giacchino, J. L., Frush, D. F., and McNamara, J. O.,** Evidence implicating dentate granule cells in lateral entorhinal cortex kindling, *Soc. Neurosci. Abstr.,* 10, 344, 1984.
7. **Willinger, M. and Margolis, D. M.,** Effect of the *weaver (wv)* mutation on cerebellar neuron differentiation. I. Qualitative observations of neuron behavior in culture, *Dev. Biol.,* 107, 156, 1985.
8. **Willinger, M. and Margolis, D. M.,** Effect of the *weaver (wv)* mutation on cerebellar neuron differentiation. II. Quantitation of neuron behavior in culture, *Dev. Biol.,* 107, 156, 173, 1985.
9. **Hatten, M. E., Liem, R. K. H., and Mason, C. A.,** *Weaver* mouse cerebellar granule neurons fail to migrate on wildtype astroglial processes "in vitro", *J. Neurosci.,* 6, 2676, 1986.
10. **Herrup, K. and Mullen, R. J.,** *Staggerer* chimeras: intrinsic nature of Purkinje cell defects and implications for normal cerebellar development, *Brain Res.,* 178, 443, 1979.
11. **Herrup, K. and Sunter, K.,** Lineage dependent and independent control of Purkinje cell number in the mammalian CNS: further quantitative studies of *Lurcher* chimeric mice, *Dev. Biol.,* 117, 1986.
12. **Mullen, R. J.,** Site of *pcd* gene action and Purkinje cell mosaicism in cerebella of chimaeric mice, *Nature,* 270, 245, 1977.
13. **Mullen, R. J. and Herrup, K.,** Chimeric analysis of mouse cerebellar mutants, in *Neurogenetics: Genetic Approaches to the Nervous System,* Breakfield, X. O., Ed., Elsevier/North Holland, New York, 1979, 271.

14. **Katsuki, M., Sato, M., Kimura, M., Yokoyama, M., Kobayashi, K., and Nomura, T.,** Conversion of normal behavior to *shiverer* by myelin basic protein antisense cDNA in transgenic mice, *Science,* 241, 593, 1988.

15. **Small, J. A., Scangos, G. A., Cork, L., Jay, G., and Khoury, G.,** The early region of the human papovavirus induces dysmyelination in transgenic mice, *Cell,* 46, 13, 1986.

16. **Sidman, R. L., Green, M. C., and Appel, S. H.,** *Catalog of the Neurological Mutants of the Mouse,* Harvard University Press, Cambridge, MA, 1965.

17. **Landis, D. M. D. and Landis, S. C.,** Several mutations in mice that affect the cerebellum, in *The Inherited Ataxias,* Kark, R. A. P., Rosenberg, R. N., and Schut, L. J., Eds., Raven Press, New York, 1978, 85.

18. **Edelman, G. M. and Chuong, C. M.,** Embryonic to adult conversion of neural cell adhesion molecules in normal and *staggerer* mice, *Proc. Natl. Acad. Sci. U.S.A.,* 79, 7036, 1982.

19. **Seiler, M. and Schwab, M. E.,** Specific retrograde transport of nerve growth factor (NGF) from neocortex to nucleus basalis in the rat, *Brain Res.,* 300, 33, 1984.

20. **Schwartz, J. P., Ghetti, B., Truex, L., and Schmidt, M. J.,** Increase of a nerve growth factor-like protein in the cerebellum of *pcd* mutant mice, *J. Neurosci. Res.,* 8, 205, 1982.

21. **Whittemore, S. R., Friedman, P. L., Larhammer, D., Persson, H., Gonzalez-Caravajal, M., and Holets, V. R.,** Rat beta-nerve growth factor sequence and site of synthesis in the adult hippocampus, *J. Neurosci. Res.,* 20, 403, 1988.

22. **Rennert, R. D. and Heinrich, G.,** Nerve growth factor mRNA in brain. Localization by *in situ* hybridization, *Biochem. Biophys, Res. Commun.,* 138, 813, 1986.

23. **Ayer-LeLivre, C., Olson, L., Ebendal, T., Sieger, A., and Persson, H.,** Expression of the beta-nerve growth factor gene in hippocampal neurons, *Science,* 240, 1339, 1988.

24. **de Blas, A. L., Vitorica, J., and Friedrich, P.,** Localization of the $GABA_A$ receptor in the rat brain with a monoclonal antibody to the 57,000 M_r peptide of the $GABA_A$ receptor/benzodiazepine receptor/Cl- channel complex, *J. Neurosci.,* 8, 602, 1988.

25. **Sequier, J. M., Richards, J. G., Malherbe, P., Price, G. W., Mathews, S., and Mohler, H.,** Mapping of brain areas containing RNA homologous to cDNAs encoding the alpha and beta subunits of the rat $GABA_A$ gamma-aminobutyrate receptor, *Proc. Natl. Acad. Sci. U.S.A.,* 85, 7815, 1988.

26. **Nishimori, T., Moskowitz, M. A., and Uhl, G. R.,** Opioid peptide gene expression in rat trigeminal nucleus caudalis neurons: normal distribution and effects of trigeminal deafferentation, *J. Comp. Neurol.,* 274, 142, 1988.

27. **Brann, M. R. and Young, S. W.,** Localization and quantitation of opsin and transducin mRNAs in bovine retina by *in situ* hybridization histochemistry, *FEBS Lett.,* 200, 275, 1986.

28. **Siegel, R. E.,** The mRNAs encoding $GABA_A$/benzodiazepine receptor subunits are localized in different cell populations of the bovine cerebellum, *Neuron,* 1, 579, 1988.

29. **Wisden, W., Morris, B. J., Darlison, M. G., Hunt, S. P., and Barnard, E. A.,** Distinct $GABA_A$ receptor a subunit mRNAs show differential patterns of expression in bovine brain, *Neuron,* 1, 937, 1988.

30. **Kreschatisky, M., et al.,** in preparation.

31. **Llinas, R. R.,** The cortex of the cerebellum, *Sci. Am.,* 232 56, 1975.

32. **Eccles, J. C., Ito, M., and Szentagothai, J.,** *The Cerebellum as a Neuronal Machine,* Springer-Verlag, New York, 1967.

33. **Hausmann, B., Mangold, U., Sievers, J., and Berry, M.,** Derivation of cerebellar Golgi neurons from the external granular layer: evidence from explantation of external granular cells *in vivo, J. Comp. Neurol.,* 232, 511, 1985.

34. **Altman, J. and Bayer, S. A.,** Embryonic development of the rat cerebellum. III. Regional differences in the time of origin, migration and settling of Purkinje cells, *J. Comp. Neurol.,* 231, 42, 1985.

35. **Inouye, M. and Murakami, U.,** Temporal and spatial patterns of Purkinje cell formation in the mouse cerebellum, *J. Comp. Neurol.,* 198, 499, 1980.

36. **Wassef, M., Zanetta, J.-P., Brehier, A., and Sotelo, C.,** Transient biochemical compartmentalization of Purkinje cells during early cerebellar development, *Dev. Biol.,* 111, 129, 1985.

37. **Wassef, M., Sotelo, C., Chelley, B., Brehier, A., and Thomasset, M.,** Cerebellar mutations affecting the postnatal survival of Purkinje cells in the mouse disclose a longitudinal pattern of differentially sensitive cells, *Dev. Biol.,* 124, 379, 1987.

38. **Herrup, K. and Mullen, R. J.,** Regional variation and absence of large neurons in the cerebellum of the *staggerer* mouse, *Brain Res.,* 172, 1, 1979.

39. **Goffinet, A. M.,** The embryonic development of the cerebellum in normal and *reeler* mutant mice, *Anat. Embryol. (Berlin),* 168, 73, 1983.

40. **Terashima, T., Inoue, K., Inoue, Y., Yokoyama, M., and Mikoshiba, K.,** Observations on the cerebellum of normal-*reeler* mutant mouse chimera, *J. Comp. Neurol.,* 252, 264, 1986.

41. **Mullen, R. J.,** Genetic dissection of the CNS with mutant-normal mouse and rat chimeras, in *Approaches to the Cell Biology of Neurons,* Cowan, W. M. and Ferrendelli, J. A., Eds., Society of Neuroscience, Bethesda, MD, 1977, 47.

42. **Blatt, G. J. and Eisenman, L. M.,** Topographic and zonal organization of the olivocerebellar projection in the *reeler* mutant mouse, *J. Comp. Neurol.,* 267, 603, 1988.

43. **Mikoshiba, K., Terada, S., Takamatsu, K., Shimai, K., and Tsukada, Y.,** Histochemical and immunohistochemical studies of the cerebellum from the *reeler* mutant mouse, *Dev. Neurosci.,* 6, 101, 1984.

44. **Mariani, J., Crepel, F., Mikoshiba, K., Changeux, J. P., and Sotelo, C.,** Anatomical, physiological and biochemical studies of the cerebellum from *reeler* mutant mouse, *Phil. Trans. R. Soc. London,* 281, 1, 1977.

45. **Shojaeian, H., Delhaye-Bouchard, N., and Mariani, J.,** Decreased number of cells in the inferior olivary nucleus of the developing *staggerer* mouse, *Dev. Brain Res.,* 21, 141, 1985.

46. **Crepel, F., DuPont, J.-L., and Gardette, R.,** Selective absence of calcium spikes in Purkinje cells of *staggerer* mutant mice in cerebellar slices maintained *in vitro, J. Physiol. (London),* 346, 111, 1984.

47. **Sidman, R. L., Lane, P. W., and Dickie, M. M.,** *Staggerer:* a new mutation in the mouse affecting the cerebellum, *Science,* 137, 610, 1962.

48. **Rezai, Z. and Yoon, C. H.,** Abnormal rate of granule cell migration in the cerebellum of *"weaver"* mutant mice, *Dev. Biol.,* 29, 17, 1972.

49. **Herrup, K. and Trenkner, E.,** Regional differences in cytoarchitecture of the *weaver* cerebellum suggest a new model for *weaver* gene action, *Neuroscience,* 23, 871, 1987.

50. **Goldowitz, D. and Mullen, R. J.,** Granule cell as a site of gene action in the *weaver* mouse cerebellum. Evidence from heterozygous *weaver* chimeras, *J. Neurosci.,* 2, 1474, 1982.

51. **Grover, B. G. and Grusser-Cornehls, U.,** Cerebellar afferents in normal and *weaver* mutant mice, *Brain Behav. Evolution,* 29, 162, 1986.

52. **Wetts, R. and Herrup, K.,** Interaction of granule, Purkinje and inferior olivary neurons in *Lurcher* chimeric mice. I. Qualitative studies, *J. Embryol. Exp. Morphol.,* 68, 87, 1982.

53. **Mullen, R. J.,** Analysis of CNS development with mutant mice and chimeras, in *Genetic Approaches to Developmental Neurobiology,* Tsukada, Y., Ed., Springer-Verlag, Berlin, 1983.

54. **Shojaeian, H., Delhaye-Bouchard, N., and Mariani, J.,** Stability of inferior olivary neurons in rodents. I. Moderate cell loss in adult *Purkinje cell degeneration* mutant mouse, *Dev. Brain Res.,* 38, 211, 1988.

55. **Caddy, K. W. T. and Biscoe, T. J.,** Structural and quantitative studies on the normal C3H and *Lurcher* mutant mouse, *Phil. Trans. R. Soc. London Ser. B.,* 287, 167, 1979.

56. **Landis, S. C.,** Ultrastructural changes in the mitochondria of cerebellar Purkinje cells of *"nervous"* mutant mice, *J. Cell Biol.,* 57, 782, 1973.

57. **Wuenschell, C. W., Fisher, R. S., Kaufman, D. L., and Tobin, A. J.,** *In situ* hybridization to localize mRNA encoding the neurotransmitter synthetic enzyme glutamate decarboxylase in mouse cerebellum, *Proc. Natl. Acad. Sci. U.S.A.,* 83, 6193, 1986.

58. **Wuenschell, C. W. and Tobin, A. J.,** The abnormal cerebellar organization of *weaver* and *reeler* mice does not affect the cellular distribution of three neuronal mRNAs, *Neuron,* 1, 805, 1988.

59. **Wuenschell, C. W. et al.,** manuscript in preparation.

60. **Parkes, C. O., Mariani, J., and Thomasset, M.,** 28K cholecalcin (CaBP) levels in abnormal cerebella: studies in mutant mice and harmaline and 3-acetylpyridine-treated rats, *Brain Res.,* 339, 265, 1985.

61. **Heinlein, U. A., Ruppert, C., and Wille, W.,** *Staggerer*-specific protein SP47: a unique species among age- and genotype-dependent cerebellar proteins, *Neurochem. Res.,* 12, 53, 1987.

62. **Frantz, G. D. et al.,** manuscript in preparation.

63. **Goldowitz, D., Barthels, D., and Wille, W.,** Localization of N-CAM mRNA in the developing mouse CNS, *Soc. Neurosci. Abstr.,* 13, 1396, 1987.

64. **Cheung, W. Y.,** Calmodulin plays a pivotal role in cellular regulation, *Science,* 207, 19, 1980.

65. **Caceres, A., Bender, P., Snavely, L., Rebhun, L. I., and Steward, O.,** Distribution and subcellular localization of calmodulin in adult and developing brain tissue, *Neuroscience,* 10, 449, 1983.

66. **Seto-Ohshima, A., Keino, H., Kitajima, S., Sano, M., and Mizutani, A.,** Developmental change of the immunoreactivity to anti calmodulin antibody in the mouse brain, *Acta Histochem. Cytochem.,* 17, 109, 1984.

67. **Branks, P. L. and Wilson, M. C.,** Patterns of gene expression in the murine brain revealed by *in situ* hybridization of brain specific mRNAs, *Brain Res.,* 387, 1, 1986.

68. **Messer, A., Wylen, E. L., Plummer-Siegard, J. A., and Wilson, M. C.,** Gene expression in developing and mutant Purkinje cells, *Soc. Neurosci. Abstr.,* 14, 1323, 1988.

69. **Messer, A., Plummer-Siegard, J., and Eisenberg, B.,** *Staggerer* mutant mouse Purkinje cells do not contain detectable calmodulin mRNA, manuscript in preparation.

70. **Bayer, S. A.,** Hippocampal region, *The Rat Nervous System,* Vol. 1, *Forebrain and Midbrain,* Paxinos, G., Ed., Academic Press, Orlando, FL, 1985, 335.

71. **Caviness, V. S., Jr. and Rakic, P.,** Mechanisms of cortical development: a view from mutations in mice, *Ann. Rev. Neurosci.,* 1, 297, 1978.

72. **Bliss, T. V. P. and Chung, S. H.,** An electrophysiological study of the hippocampus of the *'reeler'* mutant mouse, *Nature,* 252, 153, 1974.

73. **Stanfield, B. B., Caviness, V. S., Jr., and Cowan, W. M.,** The organization of certain afferents to the hippocampus and dentate gyrus in normal and *reeler* mice, *J. Comp. Neurol.,* 185, 461, 1979.

74. **Oyler, G. A., Higgins, G. A., Hart, R. A., Bloom, F. E., Battenberg, E., and Wilson, M. C.,** The sequence and characterization of SNAP-25, a neuronal specific gene, manuscript in preparation.

75. **Wilson, M.,** personal communication.

76. **Campagnoni, A. T.,** Molecular biology of myelin proteins from the central nervous system, *J. Neurochem.,* 51, 2, 1988.

77. **Lemke, G.,** Unwrapping the genes of myelin, *Neuron,* 1, 535, 1988.

78. **Sorg, B. A., Smith, M. M., and Campagnoni, A. T.,** Developmental expression of the myelin proteolipid protein and basic protein mRNAs in normal and dysmyelinating mutant mice, *J. Neurochem.,* 49, 1146, 1987.

79. **Zeller, N. K., Behar, T. N., Dubois-Delacq, M. E., and Lazzarini, R. A.,** Timely expression of myelin basic protein gene in rat brain oligodendrocytes cultured in the absence of neurons, *J. Neurosci.,* 5, 2955, 1985.

80. **Trapp, B. D., Moench, T., Pulley, M., Barbosa, E., Tennekoon, G., and Griffin, J.,** Spatial segregation of mRNA encoding myelin-specific proteins, *Proc. Natl. Acad. Sci. U.S.A.,* 84, 7773, 1987.

81. **Verity, A. N. and Campagnoni, A. T.,** Regional expression of myelin protein genes in the developing mouse brain: *in situ* hybridization studies, in press.

82. **Roach, A., Takahashi, N., Pravtcheva, D., Ruddle, F., and Hood, L.,** Chromosomal mapping of mouse myelin basic protein gene and structure and transcription of the partially deleted gene in *shiverer* mutant mice, *Cell,* 42, 149, 1985.

83. **Akowitz, A. A., Barbarese, E., Scheld, K., and Carson, J. H.,** Structure and expression of myelin basic protein gene sequences in the *mld* mutant mouse: reiteration and rearrangement of the MBP gene, *Genetics,* 116, 447, 1987.

84. **Roch, J. M., Brown-Luedi, M., Cooper, B. J., and Matthieu, J. M.,** Mice heterozygous for the *mld* mutation have intermediate levels of myelin basic protein mRNA and its translation products, *Brain Res.,* 387, 137, 1986.

85. **Nave, K.-A., Bloom, F. E., and Milner, R. J.,** A single nucleotide difference in the gene for myelin proteolipid protein defines the *jimpy* mutation in the mouse, *J. Neurochem.,* 49, 1873, 1987.

86. **Macklin, W. B., Gardinier, M. V., King, K. D., and Kampf, K.,** An AG-GG transition at a splice site in the myelin proteolipid protein gene in *jimpy* mice results in the removal of an exon, *FEBS Lett.,* 223, 417, 1987.

87. **Knapp, P. E., Skoff, R. P., and Redstone, D. W.,** Oligodendroglial cell death in *jimpy* mice: an explanation for the myelin deficit, *J. Neurosci.,* 6, 2813, 1986.

88. **Kerner, A.-L. and Carson, J. H.,** Effect of the *jimpy* mutation on expression of myelin proteins in heterozygous and hemizygous mouse brain, *J. Neurochem.,* 43, 1706, 1984.

89. **Frail, D. E. and Braun, P. E.,** Abnormal expression of the myelin-associated glycoprotein in the central nervous system of dysmyelinating mutant mice, *J. Neurochem.,* 45, 1071, 1985.

90. **Carnow, T. B., Carson, J. H., Brostoff, S. W., and Hogan, E. L.,** Myelin basic protein gene expression in q*uaking, jimpy,* and *myelin-synthesis deficient* mice, *Dev. Biol.,* 106, 38, 1984.

91. **Hogan, E. L. and Greenfield, S.,** Animal models of genetic disorders in myelin, in *Myelin,* Morell, P., Ed., Plenum Press, New York, 1984, 489.

92. **Konat, G., Trojanowska, M., Gantt, G., and Hogan, E. L.,** Expression of myelin protein genes in *quaking* mouse brain, *J. Neurosci. Res.,* 20, 19, 1988.

93. **Trapp, B. D., Small, J. A., Pulley, M., Khoury, G., and Scangos, G. A.,** Dysmyelination in transgenic mice containing JC virus early region, *Ann. Neurol.,* 23, 38, 1988.

94. **Verity, A. N. and Campagnoni, A. T.,** Myelin basic protein gene expression in the *quaking* mouse, manuscript in preparation.

95. **Coyle, J. T., Oster-Granite, M. L., Reeves, R. H., and Gearhart, J. D.,** Down Syndrome, Alzheimer's disease and the *trisomy 16* mouse, *Trends Neurosci.,* 11, 390, 1988.

96. **Saltarelli, M. D., Forloni, G. L., Oster-Granite, M. L., Gearhart, J. D. and Coyle, J. T.,** Neurochemical characterization of embryonic brain development in *Trisomy 19 (Ts19)* mice: implications of selective deficits observed for abnormal neural development in aneuploidy, *Dev. Genet.,* 8, 267, 1987.

97. **O'Hara, B. F., Reeves, R. H., Bendotti, C., Fisher, S., Capone, G. T., Coyle, J. T., Oster-Granite, M. L., and Gearhart, J. D.,** Genetic mapping and developmental expression of genes on mouse chromosome 16 in normal and aneuploid mice, *Soc. Neurosci. Abstr.,* 14, 828, 1988.

98. **Capone, G. T., Bendotti, C., O'Hara, B. F., Oster-Granite, M. L., Gearhart, J. D., Reeves, R. H., and Coyle, J. T.,** Expression of the gene coding growth associated protein-43 (GAP-43) in the brains of normal and trisomic mice, *Soc. Neurosci. Abstr.,* 14, 828, 1988.

99. **Bendotti, C., Forloni, G. L., Morgan, R. A., O'Hara, B. F., Oster-Granite, M. L., Reeves, R. H., Gearhart, J. D., and Coyle, J. T.,** Neuroanatomical localization and quantification of amyloid precursor protein mRNA by *in situ* hybridization in the brains of normal, aneuploid and lesioned mice, *Proc. Natl. Acad. Sci., U.S.A.,* 85, 3628, 1988.

100. **Oberdick, J., Levinthal, F., and Levinthal, C.,** A Purkinje cell differentiation marker shows a partial DNA sequence homology to the cellular sis/PDGF2 gene, *Neuron,* 1, 376, 1988.

101. **Ziai, M. R., Pan, Y. C.-E., Hulmes, J. D., Sangameswaran, L., and Morgan, J. I.,** Isolation, sequence and developmental profile of a brain-specific polypeptide, PEP-19, *Proc. Natl. Acad. Sci. U.S.A.,* 83, 8240, 1986.

102. **Levinthal, F., Oberdick, J., Yang, S. M., and Levinthal, C.,** Specific mRNA identified during development in mouse Purkinje cells and their dendrites, *Soc. Neurosci. Abstr.,* 13, 1706, 1987.

103. **Nordquist, D. T., Kozak, C. A., and Orr, H. T.,** cDNA cloning and characterization of three genes uniquely expressed in cerebellum by Purkinje neurons, *J. Neurosci.,* 8, 1988.

104. **Travis, G. H., Brennan, M. B., Danielson, P. E., Kozak, C. A., and Sutcliffe, J. G.,** Identification of a photoreceptor specific mRNA encoded by the gene responsible for *retinal degeneration slow (rds), Nature,* in press.

105. **Ruppert, C., Goldowitz, D., and Wille, W.,** Proto-oncogene c-myc is expressed in cerebellar neurons at different developmental stages, *EMBO J.,* 5, 1897, 1986.

106. **Wilkinson, D. G., Bailes, J. A., and McMahon, A. P.,** Expression of the proto-oncogene int-1 is restricted to specific neural cells in the developing mouse enbryo, *Cell,* 50, 79, 1987.

107. **Wilcox, J. N. and Derynck, R.,** Localization of cells synthesizing transforming growth factor-alpha mRNA in the mouse brain, *J. Neurosci.,* 8, 1901, 1988.

108. **Davis, C. A., Noble-Topham, S. E., Rossant, J., And Joyner, A. L.,** Expression of the homeobox-containing gene En-2 delineates a specific region of the developing mouse brain, *Genes Dev.,* 2, 361, 1988.

109. **Schughart, K., Utset, M. F., Awgulewitsch, A., and Ruddle, F. H.,** Structure and expression of Hox-2.2, a murine homeobox-containing gene, *Proc. Natl. Acad. Sci. U.S.A.,* 85, 5582, 1988.

110. **Rotter, A. and Frostholm, A.,** Cerebellar benzodiazepine receptors: cellular localization and consequences of neurological mutations in mice, *Brain Res.,* 444, 133, 1988.

111. **Rotter, A. and Frostholm, A.,** Cerebellar histamine H-1 receptor distribution: an autoradiographic study of Purkinje cell degeneration in *staggerer, weaver* and *reeler* mutant mouse strains, *Brain Res. Bull.,* 16, 205, 1986.

112. **Rotter, A. and Frostholm, A.,** The localization of $GABA_A$ receptors in mice with mutations affecting the structure and connectivity of the cerebellum, *Brain Res.,* 439, 236, 1988.

113. **Goffinet, A. M. and Caviness, V. S., Jr.,** Autoradiographic localization of beta 1 and alpha 1-adrenoceptors in the midbrain and forebrain of normal and *reeler* mutant mice, *Brain Res.,* 366, 193, 1986.

114. **Levitt, P., Lau, C., Pylypiw, A., and Ross, L. L.,** Central adrenergic receptor changes in the inherited noradrenergic hyperinnervated mutant mouse *tottering, Brain Res.,* 418, 1987.

115. **Lorton, D. and Davis, J. N.,** The distribution of beta 1 and beta 2-adrenergic receptors of normal and *reeler* mouse brain: an *in vitro* autoradiographic study, *Neuroscience,* 23, 199, 1987.

116. **Fisher, M. and Mullen, R. J.,** Neuronal influence on glial enzyme expression: evidence from chimeric mouse cerebellum, *Neuron,* 1, 151, 1988.

117. **Persohn, E. and Schachner, M.,** Immunoelectron microscopic localization of the neural cell adhesion molecules L1 and N-CAM during postnatal development of the mouse cerebellum, *J. Cell Biol.,* 105, 569, 1987.

118. **Faissner, A., Kruse, J., Neike, J., and Schachner, M.,** Expression of neural cell adhesion molecule L1 during development in neurological mutants and in the peripheral nervous system, *Dev. Brain Res.,* 15, 69, 1984.

119. **Zilla, P., Celio, M. R., Fasol, R., and Zenker, W.,** Ectopic parvalbumin-positive cells in the cerebellum of the adult mutant mouse *"nervous", Acta Anat.,* 124, 181, 1985.

120. **Schachner, M.,** Glial antigens and the expression of neuroglial phenotypes, *Trends Neurosci.,* 225, 1982.

121. **Levine, S. M., Seyfried, T. N., Yu, R. K., and Goldman, J. E.,** Immunocytochemical localization of GD3 ganglioside to astrocytes in murine cerebellar mutants, *Brain Res.,* 374, 260, 1986.

122. **Slemmon, J. R., Goldowitz, D., Blacher, R., and Morgan, J. I.,** Evidence for the transneuronal regulation of cerebellin biosynthesis in developing Purkinje cells, *J. Neurosci.,* 8, 4603, 1988.
123. **Ghandour, M. S., Derer, P., Labourdette, G., Delaunoy, J. P., and Langley, O. K.,** Glial cell markers in the *reeler* mutant mouse: a biochemical and immunohistological study, *J. Neurochem.,* 36, 195, 1981.
124. **Caddy, K. W. T., Patterson, D. L., and Biscoe, T. J.,** Use of the UCHT1 monoclonal antibody to explore mouse mutants and development, *Nature,* 300, 441, 1982.
125. **Weber, A. and Schachner, M.,** Development and expression of cytoplasmic antigens in Purkinje cells recognized by monoclonal antibodies: studies in neurologically mutant mice, *Cell Tissue Res.,* 227, 659, 1982.
126. **Ziai, M. R., Sangameswaran, L., Hempstead, J. L., Danho, W., and Morgan, J. I.,** An immunohistochemical analysis of the distribution of a brain-specific polypeptide, PEP-19, *J. Neurochem.,* 51, 1771, 1988.
127. **Hess, D. T. and Hess, A.,** 5′ Nucleotidase of cerebellar molecular layer: reduction in Purkinje cell-deficient mutant mice, *Dev. Brain Res.,* 29, 93, 1986.
128. **Triarhou, L. C. and Ghetti, B.,** Monoaminergic nerve terminals in the cerebellar cortex of *Purkinje cell degeneration* mutant mice: fine structural integrity and modification of cellular environs following loss of Purkinje and granule cells, *Neuroscience,* 18, 795, 1986.
129. **Felten, D. L., Felten, S. Y., Perry, K. W., Fuller, R. W., Nurnberger, J. I., and Ghetti, B.,** Noradrenergic innervation of the cerebellar cortex in normal and *Purkinje cell degeneration* mutant mice: evidence for long term survival following loss of the two major cerebellar cortical neuronal populations, *Neuroscience,* 18, 783, 1986.
130. **Kostrzewa, R. M. and Harston, C. T.,** Altered histofluorescent pattern of noradrenergic innervation of the cerebellum of the mutant mouse *Purkinje cell degeneration, Neuroscience,* 18, 809, 1986.
131. **Brugge, J. S., Lustig, A., and Messer, A.,** Changes in the pattern of expression of pp60c-src in cerebellar mutants of mice, *J. Neurosci. Res.,* 18, 532, 1987.
132. **Kaseda, Y., Ghetti, B., Low, W. C., Richter, J. A., and Simon, J. R.,** Dopamine D2 receptors increase in the dorsolateral striatum of *weaver* mutant mice, *Brain Res.,* 422, 178, 1987.
133. **Panagopoulos, N., Matsokis, N. A., and Valcana, T.,** Kinetic and pharmacologic characterization of dopamine binding in the mouse cerebellum and the effects of the *reeler* mutation, *J. Neurosci. Res.,* 19, 122, 1988.
134. **Wille, W. and Trenkner, E.,** Changes in particulate neuraminidase activity during normal and *staggerer* mutant mouse development, *J. Neurochem.,* 37, 443, 1981.
135. **Ohsugi, K., Adachi, K., and Ando, K.,** Serotonin metabolism in the CNS of cerebellar ataxic mice, *Experientia,* 42, 1245, 1986.
136. **Messer, A., Savage, M., and Carter, T. P.,** Thymidine kinase activity is reduced in the developing *staggerer* cerebellum, *J. Neurochem.,* 37, 1610, 1981.
137. **Shur, B. D.,** Galactosyltransferase defects in *reeler* mouse brain, *J. Neurochem.,* 39, 201, 1982.
138. **Wille, W., Heinlein, U. A., Spier-Michl, I., Thielsch, H., and Trenkner, E.,** Development-dependent regulation of *N*-acetyl-beta-D-hexosaminidase of cerebellum and cerebrum of normal and *staggerer* mutant mice, *J. Neurochem.,* 40, 235, 1983.
139. **Chan-Palay, V. and Palay, S. L.,** Cerebellum, in, *Encyclopedia of Neuroscience,* Vol. 1, Adelman, G., Ed., Birkhauser Boston, Cambridge, MA, 1987, 194.
140. **Teyler, T. J., Foy, M. R., Chidia, N. L., and Vardaris, R. M.,** Gonadal steroid modulation of hippocampus, in *The Hippocampus,* Vol. 3, Isaacson, R. L. and Pribram, K. H., Eds., Plenum Press, New York, 1986, chap. 10.

Chapter 5

IN SITU HYBRIDIZATION AS A MEANS OF STUDYING THE ROLE OF GROWTH FACTORS, ONCOGENES, AND PROTO-ONCOGENES IN THE NERVOUS SYSTEM

Marion Murray

TABLE OF CONTENTS

I. INTRODUCTION

The complex organization of the mammalian nervous system is achieved through interactions among neurons and between neurons and their supporting cells. These interactions may be brief, but the consequences may persist. Some are temporary, mediating a transient response to a stimulus; others may last longer and can mediate adaptive long-term changes. Some, associated wtih terminal differentiation, last for the lifetime of the cell. Determining the relation between changes in patterns of gene expression and changes in behavior of neurons and their supporting cells resulting from developmental processes or external stimuli offers new ways of investigating the organization of the nervous system. Because *in situ* hybridization allows identification of the specific cells whose gene expression is altered by some event, it is now possible to begin to examine the cellular interactions by which gene expression is regulated and thus to examine the means by which neuronal differentiation and plasticity are regulated.

In recent years, cancer biologists have uncovered properties of groups of genes whose protein products regulate developmental processes. These properties include oncogenes and their normal cellular counterparts, proto-oncogenes, and the extracellular hormones which influence them, the growth factors. Many of the proto-oncogenes, oncogenes, and growth factors that come into play in tumorigenesis are also found in high levels in the nervous system and show specific changes in levels of expression associated with developmental or other events, consistent with regulatory roles in the development and function of the nervous system.

Proto-oncogenes comprise a relatively small group of normal cellular genes which serve as substrates for many of the mutations leading to tumorigenesis. Oncogenes are the mutated or "activated" forms of the proto-oncogenes which alter the expression of proto-oncogenes or their protein products and thus transform cells by inducing a state of unregulated growth.[1] In many cases, proto-oncogenes and oncogenes share sequence homologies with known growth factors or with membrane or intracellular receptors for growth factors. Some proto-oncogenes encode growth factors or their receptors, the protein products of others may resemble, mimic, or be induced by growth factors, and some bind to DNA and can thereby directly regulate expression of other genes. The products of proto-oncogenes thus constitute steps in the pathway for transmission of extracellularly derived information to the nucleus of the cell where gene expression is regulated.

Polypeptide growth factors are products of a functional family of genes, not all of which are considered to be proto-oncogenes. Growth factors induce proliferation and differentiation in target cells, and many growth factors promote entry of resting cells into the cell cycle, a feature of importance for developing neurons and glial cells. Cells differentiate from immature progenitors, losing their capacity to divide, under the influence of growth factors, a feature which is important to postmitotic neurons and glial cells.

At present, little is known about the role of proto-oncogenes or the mechanisms underlying the effects of growth factors upon neurons and glial cells. In this chapter, studies are considered in which the cellular resolution offered by *in situ* hybridization methods has allowed insight into the mechanisms by which growth factors and proto-oncogenes may participate in the development and function of the nervous system.

II. CLASSIFICATION

Approximately 40 proto-oncogenes and oncogenes have been recognized, yet their protein products act through only a limited number of biochemical mechanisms.[2,3] Four families of proto-oncogenes are usually recognized (Table 1).

A. Growth Factors

Neurotrophic factors promote survival of neurons and regulate neurite outgrowth during development and they may continue to influence neurons in the adult. They are relatively specific; each neurotrophic factor promotes survival of certain kinds of neurons during specific stages of development. It is thought that developing neurons can compete for limited supplies of the neurotrophic factor and that success in this competition is a determinant for survival. Neurotrophic factors may also be involved in other aspects of neuronal development and maintenance, from axonal guidance to regulation of transmitter synthesis to survival of adult neurons after lesions. In addition, some growth factors regulate glial proliferation and differentiation.

Most polypeptide growth factors recognize receptors on the external surface of cells. They bind to cell surface receptors and cytoplasmic second messengers are rapidly generated which in turn stimulate those cytoplasmic and nuclear events required for cell proliferation or differentiation.

B. src Family

This is the largest and most complex of the proto-oncogene families. The genes encode proteins with shared amino acid sequences which have protein kinase activity, usually specific for tyrosine residues. Genes that encode growth factor receptors with kinase activity are thus considered members of the src family. The protein products are usually associated with the plasma membrane. c-src, itself, is the cellular homolog of the Rous sarcoma virus gene; the gene product is $pp60^{c-src}$, a phosphoprotein that is a kinase with specificity for tyrosine residues.[4]

C. ras Family

This group of proto-oncogenes encodes proteins that transmit hormonal and growth signals from the membrane to the interior of the cell. ras proteins are homologs of G-proteins, a family of guanine nucleotide binding proteins. They are located on the plasma membrane[5] and they act to transduce signals from

TABLE 1

Group	Examples expressed in nervous system	Properties of protein product
Polypeptide growth factor	NGF, IGF, EGF, Neu, IGF, FGF, Int-1, Int-2, PDGF	Promote survival, proliferation or differentiation of receptive cells. Recognize transmembrane receptors. Generally extracellular
src	src, erb-B, IGF receptor	Growth factor receptors. Protein kinases, generally tryosine specific. Located at plasma membrane or in cytoplasm
ras	ras	GTP binding proteins. Located at plasma membrane or in cytoplasm
Nuclear proto-oncogenes	myc, fos	Activate promotors or bind DNA. Located in nucleus

Modified from Varmus, 1987.

receptors on the cell membrane to cAMP and other intracellular second messenger systems. The protein product is required for progression through the cell cycle and it may therefore be involved in mitogenic signal transduction mechanisms involving other proto-oncogenes, i.e., those related to growth factors and receptors.[6]

ras proteins were first identified as products of transforming genes of murine sarcoma viruses. The ras protein, p21, is structurally and biochemically similar to G proteins and possesses GTPase activity.[7]

D. Nuclear Proto-Oncogenes

These genes are functionally related to one another by nature of their response to growth stimuli. They are rapidly and transiently induced at high levels following stimulation by a variety of events and they encode nuclear proteins which bind to DNA, a feature which enables them to control expression of other genes. Nuclear proto-oncogenes are thus in a position to couple extracellular events to long-term adaptive events upon which neuronal differentiation and plasticity are based.

The best known nuclear proto-oncogenes are c-fos and c-myc. c-fos is the cellular homolog of the oncogene v-fos carried by murine osteogenic sarcoma viruses.[8] c-myc belongs to a family of structurally related genes, including N-myc, an oncogene amplified in neuroblastomas, and L-myc, an oncogene associated with lung cancer.[9]

The evidence presently available suggests a scheme in which external signals (growth factors) bind to transmembrane receptors (e.g., protein coded by the src family) which send messages via transducers (e.g., proteins coded by the ras family) to regulate expression of nuclear proto-oncogenes (e.g., proteins coded by c-myc or c-fos). The protein products of the nuclear proto-oncogenes are translocated to the nucleus where they bind to DNA. There is at present no direct evidence supporting this progression.

In situ hybridization is now the method of choice to identify those cells which synthesize and secrete growth factors and those which bear receptors for these growth factors. It is likely that most cells constitutively express proto-oncogenes coding membrane and intracellular receptors; changes in the levels of expression of specific genes in individual cells elicited by developmental or functional events can be demonstrated using *in situ* hybridization and these studies can yield new insights into the way in which neurons and their supporting cells respond to environmental signals.

III. GROWTH FACTORS, ONCOGENES, AND PROTO-ONCOGENES IN THE NERVOUS SYSTEM

A. Growth Factors and Their Receptors
1. Nerve Growth Factor (NGF)

NGF is the growth factor with the most fully documented role in the nervous system. It is active in both central (CNS) and peripheral (PNS) nervous systems. NGF is a polypeptide secreted in limited amounts from target tissues of NGF responsive neurons. It binds to receptors and the NGF-receptor complex is internalized and conveyed to the cell body via retrograde axonal transport where it can induce changes in expression of genes related to transmitters or structural molecules. The retrograde transport of NGF accounts for the high levels of NGF in sympathetic and sensory ganglia which do not express correspondingly high levels of mRNA for NGF.[10,11]Because of the intercellular and intracellular transport of this growth factor and its receptor, the cells synthesizing NGF cannot be deduced from the location of the protein, but can only be identified using the morphological resolution provided by methods such as *in situ* hybridization.

a. Nerve Growth Factor in the Peripheral Nervous System

The role of NGF has been particularly well worked out for the PNS. NGF is required for survival,[12] neurite outgrowth, and differentiation[13] in PC12 cells, sympathetic, and some sensory neurons and it selectively induces or regulates synthesis of specific proteins, including transmitter-related enzymes[14,15] and proteins encoded by the nuclear proto-oncogenes c-myc and c-fos.[16]

Induction of gene expression by NGF: NGF plays a role in the decision of a neural crest or PC12 cell to become a neuron or a chromaffin cell. NGF promotes neuronal differentiation in PC12 cells by activating a number of specific genes,[17] including those for NGF-inducible large external glycoprotein (NILE),[18] ornithine decarboxylase,[19] cytoskeletal proteins,[13,16,20] and the nuclear proto-oncogenes c-fos and c-myc,[16,21,22] while the general level of protein synthesis remains relatively constant. NGF thus appears to act in these cells at the level of regulation of gene expression.[16]

Attempts to deduce the molecular mechanism by which NGF acts to stimulate cells to express differentiated neuronal functions led to screening of cDNA libraries from PC12 cells with adult rat nervous tissue. A clone was identified which encoded an intermediate filament protein whose mRNA is induced by NGF; the induction of this mRNA paralleled differentiation of PC12 cells.[23] Since Northern blot analysis of the distribution of the clone indicated that the mRNA was abundant in sympathetic ganglia (present in brain and adrenal tissue and undetectable elsewhere), this clone became a candidate for a neuron-specific probe important in differentiation. *In situ* hybridization of this probe with tissue sections from adult rat nervous system revealed labeling of all neurons in sympathetic (superior cervical), parasympathetic (ciliary), and dorsal root ganglia, as well as motor neurons in spinal cord and brain stem, plus a limited number of other CNS neurons, including cerebellar, lateral vestibular, rubral, segmental, reticular, and some hippocampal neurons. The majority of CNS neurons, as well as all glial and Schwann cells, were unlabeled. The uneven distribution of labeling is consistent with a direct role of NGF in the differentiation of specific responsive neurons.

A somewhat different strategy was used by Anderson and Axel[24] who isolated cDNA clones for mRNAs that are abundant in adult sympathetic but not in adrenal chromaffin cells, despite the common origin of these two cell types from the neural crest. Several clones were identified that were uniquely associated with neurons, appeared to be independently regulated, and were present in both central and peripheral neurons. Northern blots were used to identify the temporal pattern of emergence of these messages during development and *in situ* hybridization analyses were used to identify the cells expressing the messages. Several distinct patterns were observed and one of the genes was shown to be inducible by NGF. *In situ* hybridization with sections of superior cervical ganglion revealed one clone that was abundant in neural and non-neural cells in the embryo coinciding with commitment to the neural identity. Its levels then declined precipitously but it was re-expressed selectively and at high levels after birth and during differentiation and maturation of the neurons. The time course of expression indicates that this gene is subject to regulation by different mechanisms at various developmental stages.

In these studies, induction of gene expression by NGF in specific neurons is clearly implicated. Further studies using *in situ* hybridization with appropriate probes to identify the cell types participating in differentiation will be required to determine whether the induction is a result of the production of NGF by the targets, development of NGF receptors, or migration of the cells through an environment rich in NGF.

Developmental processes regulated by NGF: NGF is required for the survival and differentiation of sympathetic and some sensory neurons during restricted periods of development. It is possible to clarify the role of NGF in development, using the cellular resolution provided by *in situ* hybridization methods, by determination of (1) which cells synthesize NGF, (2) the developmental stage at which NGF synthesis begins, and (3) the stage at which neurons express NGF receptors and also become dependent on NGF.[25,26] These methods have permitted identification of neuronal proteins whose synthesis is regulated by NGF, a necessary step in understanding the mechanism of trophic actions of NGF.

The sensory neurons that innervate the skin are NGF-dependent and in adults the amount of mRNA for NGF in peripheral fields depends on the amount of sensory innervation.[27] Davies[10] and others used quantitative Northern blot analyses to show that NGF synthesis in developing skin of the whisker pad does not commence until the sensory axons from the trigeminal ganglion reach their targets. *In situ* hybridization studies showed that NGF mRNA expression is concentrated in the epithelium of developing skin, the target of these sensory neurons, at the time of innervation. The onset of NGF mRNA transcription is thus initiated by the newly arriving axons. This finding is of particular importance since it seems to rule out a major role for target derived NGF in chemoattraction of developing sensory axons to the target.[28]

To prove that the NGF mRNA is confined to non-neuronal target structures, *in situ* hybridization was used to examine primary cultures from rat superior cervical ganglia which include Schwann cells and fibroblasts as well as sympathetic neurons.[25] Labeling was restricted to non-neuronal cells. Thus, the peripheral neurons innervating these structures respond to NGF but do not synthesize NGF mRNA; the high levels of NGF present in these neurons arise from incorporation and retrograde axonal transport of NGF from non-neuronal sources.

Another population of NGF dependent neurons, sympathetic neurons, innervate the iris. The dilator region of the iris receives a dense sympathetic innervation and also expresses a high level of NGF mRNA.[29] *In situ* hybridization has shown NGF mRNA to be localized mainly to the epithelial lining covering the posterior surface of the iris, indicating that epithelial cells synthesize NGF. There was no specific labeling over the sphincter region which is innervated by parasympathetic fibers.[25] Examination of dissociated iris cells showed that virtually all cells, including epithelial cells, smooth muscle cells, fibroblasts, and Schwann cells, could be labeled by *in situ* hybridization, although at different intensities.

In addition to the target cells of the peripheral axons, the suppporting cells of the peripheral nervous system, the Schwann cells, also express NGF mRNA. This was first suggested by Rush,[30] who used immunocytochemical methods to demonstrate the presence of NGF protein in Schwann cells. Further studies have shown that levels of NGF in Schwann cells are developmentally regulated. Northern blot analyses have demonstrated that NGF mRNA levels are much higher in newborn rats than in adult sciatic nerve. Down regulation to scarcely detectable adult levels occurs within the first 3 weeks of life.[31] This greater contribution of Schwann cells to NGF levels in the developing nerve implicates Schwann cells in a neurotrophic role during axonal outgrowth.

Transsections of peripheral nerves produce a marked change in the levels of

NGF detected, indicating that denervation up regulates levels of NGF in targets and Schwann cells. Shelton and Reichert,[29] using explants of iris, described a marked induction of NGF gene expression elicited by denervation in these explant cultures. After transsection of the adult sciatic (or facial) nerve, levels of NGF mRNA increased markedly distal to the level of transsection.[31,32] *In situ* hybridization was used to demonstrate NGF mRNA localized to non-neuronal cells in transsected peripheral nerves.[25,31] Regions proximal to the transsection were unlabeled but intense labeling was associated with the neuroma-like structure containing the outgrowing neurites and their associated cells. Schwann cells were labeled in transsected sensory and sympathetic nerves and were also labeled in transsected motor nerves which do not require NGF for survival. Schwann cells thus show a denervation-induced up regulation of levels of NGF. The region of enhanced expression of mRNA for NGF is confined to regions of axonal degeneration and retreats distally as the axons regenerate. The Schwann cells can be considered to act as a substitute target which supports and perhaps elicits regenerative growth. This idea is supported by recent studies using nerve grafts which have shown an apparent requirement by both central and peripheral axons for neurotrophic molecules derived from non-neuronal sources in order to regenerate.[33,34] The very low levels of NGF normally expressed by Schwann cells in adult nerves suggest that while axons may require NGF for outgrowth, their presence suppresses the production of NGF mRNA by non-neuronal cells.[35] The initiation of NGF synthesis in Schwann cells and its relationship to the onset of naturally occurring neuronal death is not yet known. It has been suggested,[10] for example, that NGF synthesis in developing nerve could be a consequence of normal axonal death and may not play a role in regulation of neuronal survival or neurite outgrowth *in vivo*.

RNA blot analysis was used to demonstrate expression of mRNA for NGF receptor in sympathetic and dorsal root ganglia.[36] Radioautographic visualization of ^{125}I-labeled NGF revealed the stage at which cell surface receptors for NGF first appear:[26] developing sensory neurons lack NGF receptors when growing to their targets, but the receptors become expressed at the time when the fibers reach the target. NGF receptor expression in explant cultures, however, was not triggered by target contact but instead appeared to be part of a differentiation program intrinsic to the sensory neurons. Nevertheless, it seems clear that sensory neurons become responsive to NGF and dependent upon its presence at the time at which they begin expression of NGF receptors. Developing motor neurons also have receptors for NGF and internalize and transport the NGF to the cell body, but NGF does not appear to act as a trophic factor in the same way as for sensory neurons.[37] Preliminary *in situ* hybridization studies with a probe for NGF receptors have shown labeling over superior cervical ganglion neurons,[38] but the developmental expression of this gene in specific cells has not yet been described.

The scheme of production of NGF by the target cell, binding to the NGF receptors on the neuron, internalization, and transport to the cell body is incomplete. Schwann cells also express NGF receptors.[39,40] Transsection of the adult sciatic nerve which increases mRNA for NGF in Schwann cells also increases levels of mRNA for the NGF receptor in Schwann cells and in the denervated tissues.[32,39] Regeneration of the axons results in marked decreases in both NGF and NGF receptors associated with Schwann cells.[31,32] These results suggest a picture of a regenerating axon being led by an 'NGF-laden substatum', turning down the level of synthesis of both NGF and NGF receptors as it advances on its way to its proper target.

Non-neuronal cells are thus implicated as the source of the NGF provided to the axons but the way in which NGF reaches the axons is not yet known. NGF could diffuse within the target or its distribution and availability may be more restricted.[10] Diffusion, which would make NGF freely available, implies a quantitative regulation of neuronal survival. Restriction to certain target field cells would provide a more selective support. One mechanism which would restrict access is the expression of low affinity NGF receptors by the cells which synthesize NGF, so that these cells can act in an autocrine fashion, and higher affinity receptors by the cells which are NGF sensitive and which would then have an advantage in the competition for NGF.[10,35] Proof of some of these hypotheses will require demonstration of the synthesis of NGF receptors by non-neuronal cells at the appropriate times using *in situ* hybridization.

b. Nerve Growth Factor in the CNS

NGF is also present in the CNS. Northern blot analysis[27] and immunocyto-chemical[41] studies have shown substantial levels of NGF in developing and adult CNS. By analogy with the peripheral nervous system, it might be expected that NGF would act on central catecholaminergic neurons, e.g., locus coeruleus and substantia nigra, but perhaps not cholinergic neurons, since cholinergic motor neurons appear not to be NGF dependent. Noradrenergic and dopaminergic neurons, however, do not show specific retrograde axonal transport of NGF[42] or respond to NGF by induction of tyrosine hydroxylase.[43] Instead, NGF-responsive neurons are found in central cholinergic systems, including hippocampus, neocortex, olfactory bulb, the septal nuclei, the nuclei of the diagonal band of Broca, and nucleus basalis of Meynert.[42,43]

In situ hybridization studies are required to learn the source of NGF synthesis in the CNS. Again by analogy with the PNS, one might expect labeling of glial cells. *In situ* hybridization, however, has shown specific labeling of granular neurons in the dentate gyrus, pyramidal neurons in the hippocampus[44-47] (Figure 1), layer II of the entorhinal cortex,[47] and olfactory cortex.[46] Surprisingly, so far there has been no detection of mRNA for NGF reported in non-neuronal cells *in vivo,* despite demonstration *in vitro* of production of NGF and other growth factors by astrocytes[48-51] and microglia.[52] Most NGF secretion by astrocytes in culture appears, however, to be regulated in a growth dependent fashion. A glial contribution to the NGF present in the CNS may be confined largely to mitotic astrocytes during development and perhaps also after injury.[53] At present, the

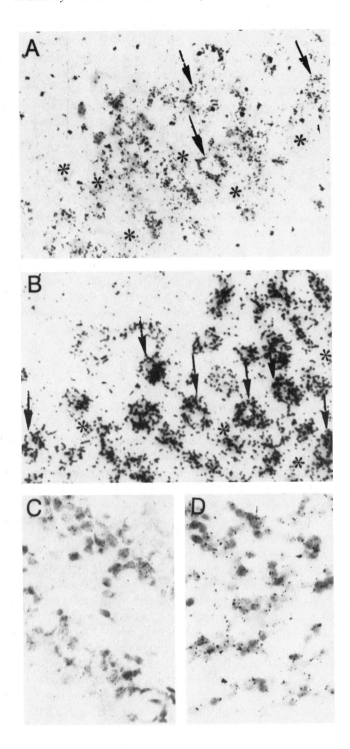

primary source of NGF in the CNS appears to be target cells in the CNS, as in the PNS, but the targets for central neurons are other neurons, while peripheral targets are non-neuronal cells.

NGF receptor mRNA has been detected in the basal forebrain in the CNS using Northern blot analysis.[36] NGF receptors have been identified immunocytochemically[54] in the basal forebrain, in the vasculature and Muller cells of the retina, and in fetal material, in the optic nerve and the cerebellum. Cells expressing mRNA for the NGF receptor in the CNS have recently been identified using *in situ* hybridization.[38] The most intensely labeled neurons were found in the medial septum, diagonal band, and nucleus basalis of Meynert, corresponding to regions showing immunocytochemical labeling, and indicating that the cells that express the mRNA for NGF receptor also manufacture the receptors.

NGF is transported retrogradely in these cholinergic neurons *in vivo*.[42,43] Quantitative Northern blot analyses indicated that NGF mRNA was predominantly found in the hippocampus and cortex, smaller amounts were present in the cerebellum and striatum but none was detected in septum.[29,50,55] The presence of NGF and mRNA for NGF receptors in the septum, but not mRNA for NGF, is consistent with a retrograde transport of NGF from target tissues in the hippocampus to septal neurons, a situation analogous to the PNS. Similarly, transsection of the septal projection to the hippocampus leads to a transient increase in hippocampal NGF content attributable to an interruption of the retrograde transport of newly synthesized NGF.[51] The effects of lesions on regulation of mRNA levels for NGF and its receptors in the CNS are as yet poorly understood but seem to differ from the peripheral effects. Cultured glial cells do not normally express NGF receptors[40] nor is there evidence for induction of NGF receptors in CNS glia in response to axotomy.[56]

NGF is reported to stimulate synthesis of choline acetyltransferase (ChAT), a synthetic enzyme for acetylcholine, in hippocampus and striatum and to increase mRNA for NGF receptors in septum.[57] NGF may also provide trophic support for these neurons. Fimbrial section results in death of the cholinergic septo-hippocampal neurons but it has been reported that the neurons can be rescued by NGF administration *in vivo*.[58,59] In these studies, however, ChAT containing neurons were counted rather than the total number of surviving neurons, so it is possible that absence of NGF modifies expression of transmitter related genes rather than causes death of cholinergic neurons.

The available information is thus consistent with a role for NGF in the CNS which is similar in many respects to that in the PNS. NGF is synthesized in the target cells of sympathetic and some dorsal root neurons in the PNS,[10,25] and in

FIGURE 1. Identification of NGF-producing cells in rat hippocampus. (A) Labeled pyramidal cells in the CA3 area (X280). (B) Granular layer of dentate gyrus (X440). (C) No labeling of hippocampal cells with control plasmid probe (X280). (D) No labeling on cerebellar granular layer (X280). Arrows indicate labeled cells, asterisks indicate localization of unlabeled cells. (Reproduced from Ayer-LeLievie et al.,[45] with permission.)

the target neurons of some cholinergic neurons in the CNS.[44-47] Survival of NGF-dependent neurons in the PNS[12] and perhaps in the CNS[58,59] appears to be dependent upon receptor mediated uptake and retrograde transport of the NGF to the cell body. NGF induces tyrosine hydroxylase and dopamine β-hydroxylase in sympathetic neurons, regulates levels of substance P and CGRP in developing and adult dorsal root ganglion cells, and induces ChAT in central neurons.[14,15,60,61]

2. Other Growth Factors

Lindsay[48] showed that astrocytes promote survival of both NGF sensitive and NGF independent neurons *in vitro*. Molecules analogous to NGF may thus be essential for growth and survival of many or most neurons, and glial cells may be the source of some of these survival factors. Many growth factors active in non-neural tissues are also present in the CNS. For some of these growth factors, detection by *in situ* hybridization has already begun to provide a way of investigating their role in the development and function of the nervous system.

a. Insulin and Insulin-like Growth Factors

Insulin and its homologs, the insulin-like growth factors (IGFs) or somatomedins, belong to a family of neuritogenic polypeptides which act as growth factors and neuromodulators, share some properties with NGF, and may play a role in nerve regeneration.[62-66] Two insulin-like growth factors have been identified, IGF-I and IGF-II. They are related peptides, the products of a single gene, share structural homology with proinsulin, and act as growth promotors for cells in developing and adult animals.

These factors, like NGF, can stimulate increased expression of the mRNAs for cytoskeletal elements,[65,67] are permissive for the induction of neurites by NGF in several cell lines,[66] and stimulate neurite outgrowth in spinal neurons.[68] They can also act as survival factors for cortical neurons, where NGF has no effect.[63]

Insulin and the IGFs are expressed at higher levels in fetal brain and then down regulated in the adult, suggesting a role in normal developmental processes.[69-73] Insulin and the IGFs play a role in regulation of the mitotic cycle in cultured rat sympathetic neuroblasts[74] and IGF-I has been shown to be synthesized by fetal rat astrocytes and to act as a growth promoter for astrocytes by activation of IGF receptors.[70] In this case, IGF-I appears to act in an autocrine fashion to stimulate astrocytic growth during normal brain development.[70,75] *In situ* hybridization has now been used to show that insulin induces mRNA for GFAP and an increased level of GFAP immunoreactivity in astroglia in culture.[76] Insulin related peptides may thus directly influence proliferation and differentiation of both neurons and glial cells.

Insulin is present in the brain, but most is probably synthesized elsewhere.[70] Insulin mRNA has been detected with *in situ* hybridization only in the periventricular hypothalamic cells[79] and not in quantities sufficient to account for the amount of insulin present in the brain.[70] Immunocytochemical studies show IGF-I to be present in spinal motor neurons, autonomic neurons, and Schwann cells, but not in glial cells[79] while IGF-II is present in these regions and also found in muscle.[62] Northern blot analyses have shown mRNAs for both IGF-I and IGF-II to be present in fetal and adult brain. IGF-I mRNA is abundant in spinal cord, cerebellum, midbrain, and olfactory bulb while IGF-II mRNA is more enriched in midbrain and striatum,[72] and hypothalamus.[79] The sites of synthesis of IGF-I and II have not yet been identified.

Sciatic nerve transsection up regulates levels of both IGF-I and IGF-II in the denervated muscle and in the portion of the nerve distal to the transsection. IGFs may thus play a role in regenerative processes, perhaps resembling the role of NGF.[73,80]

The effects of these somatomedins are mediated by two types of receptors; type I is a tyrosine specific protein kinase while the type II receptor is not. Immunocytochemical and receptor binding studies show both receptors to be present broadly within adult rat brain, including hippocampus, dentate gyrus, amygdala, and hypothalamus and to be found on both glial cells and neurons.[73,81] There is evidence for retrograde transport of IGF-I in the sciatic nerve of adult rats, presumably by binding to the receptor and being internalized in a manner resembling the NGF-receptor complex.[78]

It seems clear that insulin and the IGFs can exert trophic effects on both glial cells and neurons and that some of the actions resemble those of NGF. The sources of these trophic factors may be multiple; the cells synthesizing the insulin and IGFs found in the brain need to be identified before the roles of this family of trophic agents can be clarified. Many neurons with IGF immunoreactivity are not labeled using *in situ* hybridization. Since IGFs are transported in a retrograde manner in sciatic nerves, are present in motor neurons and since in the adult the amount of IGF appears to be regulated by innervation,[62,63] peripheral non-neural targets are candidates for the source of some of the IGFs in the CNS. Furthermore insulin can cross the blood brain barrier and can thus enter the brain from that route.[73] It is likely that much of the insulin present in the CNS is produced elsewhere and is concentrated in brain.[73]

b. Epidermal Growth Factor (EGF)

EGF is a potent mitogen for a number of cell types[82] including astrocytes.[83] EGF can also elicit some, but not all, of the responses elicited by NGF in neurons, implying a role in neuronal differentiation. Expression of c-fos and c-myc mRNA in PC12 and other cell lines is induced by EGF.[16,84] Addition of EGF to cultures of rat telencephalic neurons promotes survival of cells with neuronal phenotype, immunoreactive characteristics of neurons, and increases in the number and degree of branching of neurites.[85] Thus there is evidence that EGF acts upon both glial cells and neurons.

mRNA for EGF, detected by Northern blot analyses and SI nuclease protection assays, is present in very low levels in whole adult mouse brain[86] and is detectable in striatum and cerebellum[87] but thus far no *in situ* hybridization studies indicating the cells of origin of EGF have appeared. EGF immunoreactivity, however, is found in high concentrations in midbrain and forebrain, including most of the extrapyramidal structures.[88] The EGF immunoreactivity was visualized in 14-d-old or older rats, and apparently localized to fibers and terminals, a pattern of distribution more consistent with protein transport than protein synthesis. The location and the age dependent expression suggest that EGF is present in postmitotic neurons rather than glia and that it might have a role as a neurotransmitter-modulator rather than as a mitogen.[88]

Several EGF-related factors have been identified. Two forms of transforming growth factors (TGFs) are recognized, TGF-α and TGF-β. The product of the oncogene TGF-α shares structural homologies to EGF[89] and it binds to the EGF receptor.[90] The mRNA for TGF-α can therefore be used as a marker which may indicate sites of EGF synthesis. *In situ* hybridization using TGF-α mRNA has identified cells, primarily neurons, in caudate, dentate gyrus, anterior olfactory nuclei, and mitral cells in the olfactory bulb.[91] There are thus some similarities between the distribution of cells containing EGF immunoreactivity and those expressing TGF-α message. Cells expressing TGF-β mRNA have not been detected in brains of adult rat brain or in frog embryos.[92]

Several proteins immunologically related to EGF receptor have been identified. Immunocytochemical studies show that EGF receptors can first be detected in both neurons and glial cells in the second postnatal week. Staining becomes more intense in the next weeks and then falls to very low adult levels by day 30.[93] The time course of appearance of immunoreactivity to EGF and EGF receptor is therefore closely parallel, consistent with a coordinated regulation.

EGF receptor immunoreactivity is greatly enhanced in reactive astrocytes.[94] An astrocyte mitogen inhibitor with similarity to EGF receptor has been identified.[95] Following injury to the brain, the levels of the inhibitor fall and EGF receptor expression is enhanced, suggesting that the EGF receptors on astrocytes may contribute to the onset of reactive gliosis. The inhibitor is thought to be synthesized by astrocytes and thus may act in an autocrine fashion.[95] EGF receptors are also reported to be amplified in human brain tumors of glial origin.[96] Direct demonstration of synthesis of EGF receptors by astrocytes is lacking at present.

The protein product of another oncogene, v-erb-B, which encodes the transforming protein of the avian erthyroblastosis virus (AEV), has been shown to have close similarity to the EGF receptor.[97] erb-B has been reported to be detected in adult cerebellum using *in situ* hybridization.[98] Another recently discovered oncogene, neu, encodes a protein, p185, structurally related to but distinct from c-erb-B and thus may encode for a receptor for an unidentified growth factor, similar to EGF.[99,100] neu is expressed in glioblastoma and neuroblastoma cell lines[99] and immunocytochemical studies indicate that the neu gene is also expressed in E-14 and E-16 nervous system but in the adult, neu is expressed only in epithelium. Neu is thus expressed in a tissue and stage-specific manner consistent with a role in growth and development for a hormone related to EGF.[100]

EGF, its receptor and related proteins are therefore functionally implicated for both neurons and glia. The time course of development of immunoreactivity suggests that EGF may play an important role in post-mitotic neuronal-glial interactions. The sites of synthesis of EGF and its receptor need to be determined in both developing and adult brain and after injury in order to define the role of this growth factor.

c. Fibroblast Growth Factor (FGF)

Two forms of FGF, acidic or astrocytic growth factor 1 (aFGF), and basic or astrocytic growth factor 2 (bFGF), have been identified. They are closely related (53% homologous), but are products of separate genes. FGFs are mitogens which stimulate proliferation of a broad range of cells, including most mesodermal cells, astrocytes,[101-105] oligodendroglia,[103-106] endothelial cells,[105] and PC12 cells.[107] FGF can synchronize entry of cultured astrocytes into the cell cycle[102] and, since it is found in lesion sites following injury to the brain,[94,108] FGF is another candidate as a factor regulating glial response to brain injury.

Both forms of FGF are trophic factors for selected populations of neurons, although bFGF is more potent than aFGF. They both promote neuronal survival and neurite outgrowth.[85,105,109-111] Both forms may influence directly both neurons and glia[105] and thus both forms are candidates as true neurotrophic factors. FGF levels in rat brain increase markedly during the first month of life, the time of glial and endothelial cell proliferation, and remain high in the adult brain, implying some role in neuronal maintenance.[112] Immunocytochemical studies have shown FGF to be localized in neurons, primarily in cortex.[108-110] After injury, however, bFGF immunoreactivity is enhanced in cells resembling reactive astrocytes.[108] The distribution of binding sites for bFGF suggest that receptors are present in cerebellum, hippocampus and cortex.[113] FGF is also internalized after binding to neurons, a prerequisite for axonal transport.[105]

Another growth factor, brain derived growth factor (BDGF) shares biochemical properties with aFGF and binds to the same receptor as bFGF.[114] BDGF has been localized immunocytochemically to neurons and there is also evidence that it acts as a potent chemoattractant and mitogen for astrocytes during normal development and following injury to brain.[114,115]

Two proto-oncogenes have been identified which appear to code for growth factors, at least one of which may be related to FGFs. int-1 and int-2 are two independent genes which respond similarly to proviral insertion mutations in mammary carcinomas induced by mouse mammillary tumors but are found on different chromosomes and may have different functions.[116] Both int-1 and int-2 are expressed in development, although at different times, and neither are expressed in the adult. Proteins int-1 and int-2 may therefore play central roles in events taking place at specific times in embryogenesis.[116] With *in situ* hybridization int-1 has been shown to be restricted to neural plate cells; it is expressed in the spinal cord and in a complex pattern in the dorsal diencephalon, mesencephalon, and hindbrain in the mouse embryo during stages 9 through 15 and thus represents an extreme example of a developmentally regulated gene. The gene product of int-1 has not been further identified but it has been suggested that it may be secreted, and act as a growth factor for nearby cells. The protein product of the second oncogene, int-2, has been shown to be closely related to bFGF.[117] The predicted protein product of int-2 is larger than FGF so the two proteins could be expected to behave somewhat differently.[118,119] *In situ* hybridization studies have shown that int-2 expression is highly restricted (7.5 to 9.5 d postconception in the mouse) and localized to neuroblastic layers of retina and cerebellum and cells adjacent to the developing otocysts whose development it is in a position to influence.[119,120]

FGF and its relatives thus appear to have direct trophic effects on neurons and to stimulate proliferation in glial cells in development and in the adult. The sites of synthesis are known only for the proto-oncogene forms and those seem strictly limited developmentally; the pattern of immunocytochemical localization of the FGFs, however, suggest a continued role in the adult brain.

d. Platelet Derived Growth Factor (PDGF)

PDGF is the major mitogen present in serum and a chemoattractant which may be involved in tissue repair processes. PDGF exists as a dimer whose A and B chains are products of different genes.[121] Northern blot analysis of mRNAs encoding the PDGF A chain showed expression in primary cultures of type-I astrocytes and in neonatal rat brain.[122] PDGF has also been shown to be a potent mitogen for the O-2A progenitor cells.[123] The time course of expression of PDGF in the developing rat optic nerve suggests that type-I astrocytes may secrete PDGF which regulates proliferation and differentiation of O-2A progenitors. Support for this hypothesis was provided by application of *in situ* hybridization methods to demonstrate PDGF mRNA for the A chain, but not the B chain, associated with GFAP-positive cells in developing optic nerve at the time when the O-2A progenitor cells are proliferating.[123] These findings suggest that ho-modimers of the A chain of PDGF represent the major form of PDGF in nerve and brain and that PDGF secreted by type-I astrocytes stimulates proliferation of O-2A progenitors.[123] PDGF is also expressed at high levels in human gliomas, which is consistent with a role for PDGF in control of normal glial cell growth.[121]

The B chain of PDGF is structurally virtually identical to the protein product p28^{sis}of the v-sis oncogene of simian sarcoma virus[124,125] and this oncogene has been detected in adult cerebellum using *in situ* hybridization.[98]

PDGF receptors are restricted to cells of mesenchymal and glial origin. O-2A progenitor cells appear to possess PDGF receptors and to mediate the mitogenic response of PDGF secreted by type-1 astrocytes.[122,123,126]

PDGF and its receptors are thus important in normal gliogenesis. The extent to which PDGF in brain arises from glial sources and from nonbrain sources remains to be determined.

B. Other Proto-Oncogenes

1. src Family

This group of genes encodes receptors for growth factors. Infection of PC12 cells by retroviruses carrying the v-src oncogene induces neurite outgrowth by these cells in a way that resembles the actions of NGF, suggesting that in this cell type differentiation is under the control of the src gene.[127]

In the nervous system, c-src expression is developmentally regulated and shows clear stage and cell-type specificity.[128-131] Immunocytochemical localization of the gene product, pp60[c-src], in developing chick brain revealed an early wave of expression during neural tube formation, followed by cessation of expression, and then a second abrupt increase during neuronal differentiation.[128,131,132] The pattern suggests interesting roles, first for neuronal commitment, and again later in development, associated with differentiation.[132] The localization of the c-src gene product to growth cones of developing neurons is particularly intriguing as it may suggest a role in motility or adhesion.[133] Immunocytochemical studies have shown the c-src protein is widely distributed in both nerve cells and terminals in postmitotic retinal, cortical, cerebellar, and dorsal root ganglion neurons,[129-132,134-136] implying a role in the differentiated neuron, perhaps related to neurotransmission or synaptic plasticity. Both astrocytes and neurons express c-src in culture[137] although neurons are more highly enriched than astrocytes.

Most of the information on src expression comes from immunocytochemical and biochemical studies. Some preliminary reports from in situ hybridization studies of src expression in the nervous system have appeared. Labeling of c-src mRNA in embryonic chick neural retina has been described as being confined to amacrine and retinal ganglion cell bodies, at a time when these neuronal cells have ceased mitosis and are undergoing terminal differentiation and neurite extension.[138] In another in situ hybridization study, labeled mRNA for src has been mapped in developing and adult rat brain. C-src+, a brain specific form of the src gene product, was localized to both neurons and glial cells, and after lesions, src+ levels were found to be enhanced in glial cells. The distribution of c-src+ message showed marked regional variations; highest levels were found in cerebellar granule cells, hippocampal pyramidal layer, dentate gyrus, and in olfactory bulb and elsewhere at lower levels in the adult rat brain.[139,140] The expression of src in the adult brain suggests a general role for src expression in adult neurons, perhaps in maintenance of the differentiated neuronal state; the high levels of expression in hippocampus and olfactory bulb suggest a role in neuronal plasticity.

Another member of the src family is also expressed in brain. Erb-a, which shares homologies with steroid and thyroid hormone receptors,[141] has been labeled using in situ hybridization in the adult cerebellum of the rat.[98]

The src family has thus been implicated in developing neurons and glial cells, as would be predicted for genes encoding receptors for growth factors. There is also evidence indicating that src expression remains elevated in the adult nervous system, particularly those systems which are characterized by continued plasticity in the adult, and that its expression in glial cells may be enhanced after injury.

2. ras Family

This family of genes encodes proteins structurally related to G proteins, which hydrolyze GTP and which can thus function as signal transducing proteins, particularly with respect to growth factors.[7,142,143] The ras oncogenes are the most commonly found oncogenes isolated from human tumors and the ras product is the first oncogene protein whose structure has been determined.[142] Nevertheless, there have been only a few reports of ras expression in the nervous system.

Introducing ras oncogenes or ras protein into PC12 cells induces differentiation, including ornithine decarboxylase induction[144] and neurite outgrowth[144-146] with a latency shorter than for induction by either NGF or introduction of v-src. *In vivo* studies have shown expression of the ras protein by immunocytochemical staining in virtually all Aplysia neurons, both differentiating and mature.[5] Foote[98] reported ras expression using *in situ* hybridization in the developing cerebellum.

3. Nuclear Oncogenes
a. c-myc

The protein product of c-myc is localized in the nucleus, binds DNA, and thus is implicated in transcriptional control of other cellular genes.[147] Expression of the c-myc gene is linked to the action of growth factors and, in proliferating cells, with entry of cells into the cell cycle. C-myc mRNA levels increase after treatment of various cell lines with NGF,[16] EGF,[16,84] FGF, and PDGF[147] while transfection of PC12 cells with c-myc oncogenes blocks the differentiation of these cells by NGF and instead stimulates proliferation by these cells.[148] C-myc expression is thus implicated in at least one part of the pathway linking stimulation by some growth factors to proliferation and therefore represents an intracellular receptor for growth factor signals.

C-myc is developmentally regulated in the nervous system. It is normally expressed in dividing cells. An *in situ* hybridization study has shown that all actively dividing neuroblastoma cells show strong radioautographic signals for c-myc while treatment with agents that inhibit cell growth decreases myc mRNA levels.[149] In brain, Northern blot analyses show c-myc to be maximally expressed during early development, declining postnatally, but still expressed constitutively in adult brain.[150-152] C-myc expression is also enhanced in glial cells in regenerating optic nerves of adult fish after injury.[153]

Elevated c-myc levels are not necessarily exclusively related to proliferation.[154] This has been best shown using *in situ* hybridization to identify cells expressing c-myc in the postnatal cerebellum.[155] SI protection assays showed myc transcripts to be very high in late embryonic stages, to decrease markedly at birth, then 1 week later, to increase again, and finally to decline to low adult levels. *In situ* hybridization showed labeling in the superficial granular layer, which contains mitotically active cells but no migrating or differentiating granule cells. Purkinje cells, however, were unlabeled at the time when they were proliferating, while they were labeled at stages when they are differentiating and receiving synaptic contacts from granule cells. In the hippocampus, both proliferating granule cells and postmitotic pyramidal cells were labeled. Thus, c-myc expression in these two structures is elevated in some populations of proliferating cells and in other populations of differentiating cells (Figure 2).

b. c-fos

The protein product of the c-fos gene, like other nuclear oncogenes, binds to DNA and thus may regulate transcription of other genes.[156-158] The basal expression of the c-fos gene is very low in most cell types,[8] but a remarkable feature of this gene is its rapid and transient induction in response to a wide variety of stimuli and agents. This property suggests that c-fos may be a marker for a set of rapidly induced "cellular immediate-early" genes whose function is to couple membrane events to changes in gene expression.[8,159] This is of particular importance in the brain where c-fos expression has become a candidate for a determinant of cellular responses which could regulate neuronal plasticity.[154,160]

The fos gene is transiently inducible in a variety of cell types, including neurons, although perhaps not glial cells,[161,162,164,165] by agents which stimulate or mimic second messengers or mitogens. For example, either NGF or EGF administration[16,22,25,84,154] or injection of ras protein[6] can stimulate fos transcription in several cell lines.

Induction of c-fos is associated with differentiation as well as mitosis. C-fos expression is enhanced during development, although it is temporally linked to stages of differentiation rather than to periods of the most intense mitosis.[166] In mouse brain, Northern blot analysis shows that expression of c-fos is virtually undetectable before birth but there is a burst of expression at birth which increases over the next 5 d and subsequently returns to low levels.[152]

The protein product of c-fos can be visualized immunocytochemically in the nucleus of neurons in many regions of the adult brain, including cortex, pyramidal cells in areas CA1 and CA3, granule cells of the dentate gyrus, neurons in amygdala, striatum, piriform cortex, and cerebellum, and as well as other areas, but not in glial cells.[161,162] A variety of studies have now demonstrated that fos expression in neurons is increased dramatically and rapidly by a variety of stimuli, including repetitive firing associated with seizures,[162,167-171] increased activity associated with kindling,[163] morphine administration,[172] cortical stimulation,[173] and somatic sensory stimulation.[165] In some of these more complex experimental paradigms, it seems clear that fos expression is not enhanced equally in all stimulated neurons;[165,173] fos expression therefore appears to be regulated by events in addition to functional stimulation. Of particular interest, however, are the observations indicating that treatments leading to dehydration elicit increased fos immunoreactivity in the supraoptic and the

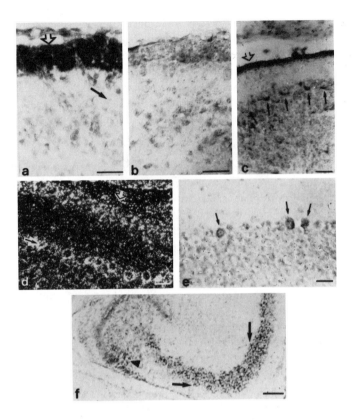

FIGURE 2. *In situ* detection of c-myc expression during postnatal cerebellar development. Biotin-probe used on tissue in a, b, c, e, f. ³H-probe used on tissue in d. (a) P3 cerebellum has reaction product associated with the external granule layer (open arrow), which is composed of mitotically active cells. Purkinje cells (arrow) have little if any expression of c-myc at this stage of development. (b) Adjacent section pretreated with ribonuclease followed by normal hybridization-protocol. (c) P10 cerebellum has reaction product localized over the cells in the germinal matrix zone of the external granule layer (open arrow) but not the mitotically quiescent deeper zone. Hybridization signal is also apparent over the cytoplasm of the differentiating Purkinje cells (arrows). (d) Cerebellar section as in c. Darkfield. Silver grains are localized at the same position as the reaction product in c (open arrow marks the matrix zone of the external granule layer; the black arrow indicates the Purkinje cell layer). (e). Adult cerebellum has very little signal with c-myc *in situ* hybridization. Some Purkinje cells (arrow) display a light staining for the probe. (e) P3 hippocampal formation displays dense staining in the pyramidal cells of regio inferior (arrows) and also some reaction product in the granule cell proliferative region (arrowhead). Scale bars for a through e = 20; f = 100 μm. (Reproduced from Rupert et al.,[55] with permission.)

paraventricular nuclei of the hypothalamus of rats[173] and also an increased expression of vasopressin mRNA over the same nuclei, as shown with *in situ* hybridization.[174,175] It still remains to be determined whether the expression of the fos gene which is elicited by the stimulus, precedes and regulates the gene encoding vasopressin. This hypothalamic system does provide an opportunity to examine proto-oncogene regulation of gene expression in a defined neuronal system resulting from adequate physiological stimulation.

The induction of c-fos after seizures shown with *in situ* hybridization is also of particular interest since the time course of induction of this gene can be followed using *in situ* hybridization in structures of the forebrain which are recruited to seizure activity.[170] *In situ* hybridization studies have shown that within 15 min of seizure induction, c-fos hybridization was increased exclusively within the dentate gyrus. At 3 h, increased labeling extended to the dentate gyrus, the stratum granulosum and layer II of piriform and entorhinal cortices, as well as in pyramidal cells throughout the hippocampus, and in neurons in the olfactory tubercle and anterior olfactory nucleus. By 6 h, the signal declined over the stratum granulosum but remained elevated in neurons in the entorhinal cortex. This study indicates not only a rapid induction of fos but also a temporal spread related to the spread of seizure activity in the brain. The seizure activity inducing c-fos also causes changes in levels of expression of neurotransmitters and their receptors which can be visualized using *in situ* hybridization.[168-171] These preliminary experiments are consistent with the suggestion that the c-fos product could mediate the increase in neurotransmitter gene expression, stimulated by increased neural activity associated with seizures.[169-171,176]

Fos induction has also been reported to be associated with injury. Stab injuries to cortex induce c-fos proteins in neurons and also in glial cells in adult rat brains.[163] Induction of c-fos in glial cells also occurs in crushed, regenerating goldfish optic nerves.[153] Finally, *in situ* hybridization has been used to demonstrate a dramatic increase in labeling of neurons following transient cerebral ischemia in caudate-putamen, CA1 of hippocampus, and neurons in the hilus of the dentate gyrus.[177]

C-fos expression therefore appears to be enhanced during development and differentiation of the brain, in response to injury, and selectively in response to increased functional activity. The use of *in situ* hybridization to show selective enhancement of expression among neurons and the time course of changes in expression offers a particularly exciting opportunity to investigate the role of physiological events in the regulation of gene expression.

IV. SUMMARY AND CONCLUSIONS

The application of *in situ* hybridization methods to the nervous system is very recent and the full potential of this approach has yet to be realized. The availability of probes for NGF and its receptor has permitted a comparison of the role of NGF in the PNS and CNS and a clear delineation of the differences and similarities in the ways in which this growth factor is used in the PNS and CNS. As probes become available for the other growth factors active in the nervous system, we can expect rapid advances in identifying the cells which secrete specific growth factors and those which bear receptors for them. This information can be expected to have considerable bearing on our understanding of the way in which the complex and specific connections are formed during development and regeneration and how they are maintained within the adult nervous system.

At present rather few experimental studies have been carried out using probes for proto-oncogenes. The development of additional probes and of quantitative methods for assessing changes in constitutive expression of proto-oncogenes at the cellular level should permit evaluation of ways in which external stimuli can modify gene expression in cells of the developing and adult nervous system. This information should contribute in a direct way to understanding developmental mechanisms, the problems of information storage within the nervous system, and the nature of adaptive neuronal plasticity.

REFERENCES

1. **Varmus, H.,** Cellular and viral oncogenes, in *Molecular Basis of Blood Diseases,* Stamatoyannopoulos, G., Ed., Saunders, Philadelphia, 1987, 271.
2. **Weinberg, R. A.,** The action of oncogenes on the cytoplasm and nucleus, *Science,* 230, 770, 1985.
3. **Bishop, J. M.,** The molecular genetics of cancer, *Science,* 238, 305, 1987.
4. **Hunter, T. and Sefton, B.,** Transforming gene product of Rous sarcoma virus phosphorylates tyrosine, *Proc. Natl. Acad. Sci. U.S.A.,* 77, 1311, 1980.
5. **Swanson, M. E., Elste, A. M., Greenberg, S. M., Schwartz, J. H., Aldrich, T. H., and Furth, M. E.,** Abundant expression of ras proteins in Aplysia neurons, *J. Cell Biol.,* 103, 485, 1986.
6. **Stacey, D. W., Watson, T., Kung, H. F., and Curran, T.,** Microinjection of transforming ras protein induces c-fos expression, *Molec. Cell Biol.,* 7, 523, 1987.
7. **Jurnak, F.,** The three-dimensional structure of c-H-ras p21: implications for oncogene and G protein studies, *TIBS,* 13, 195, 1988.
8. **Cohen, D. R. and Curran, T.,** The structure and function of the fos proto-oncogene, *Crit. Rev. Oncogenes,* in press.
9. **de Pinho, R., Mitsock, L., Hatton, K., Ferrier, P., Zimmerman, K., Legouy, E., Tesfaye, A., Collum, R., Yancopoulos, G., Nisen, P., Kriz, R., and Alt, F.,** Myc family of cellular oncogenes, *J. Cell. Biochem.,* 33, 257, 1987.
10. **Davies, A. M.,** Role of neurotrophic factors in development, *TIG,* 4, 139, 1988.
11. **Heumann, R., Korsching, S., Scott, J., and Thoenen, H.,** Relationship between levels of nerve growth factor (NGF) and its messenger RNAs in sympathetic ganglia and peripheral target tissue, *EMBO J.,* 3, 3138, 1984.
12. **Levi-Montalcini, R. and Angelotti, P. U.,** Nerve growth factor, *Physiol. Rev.,* 48, 534, 1968.
13. **Drubin, D. G., Feinstein, S. C., Shooter, E. M., and Kirschner, M. W.,** Nerve growth factor-induced neurite outgrowth in PC12 cells involves the coordinate induction of microtubule asembly and assembly-promoting factors, *J. Cell Biol.,* 101, 1799, 1985.

14. **Thoenen, H. and Barde, U. A.,** Physiology of nerve growth factor, *Ann. Rev. Physiol.,* 60, 284, 1980.

15. **Lindsay, R. M. and Harmar, A. J.,** NGF regulates the expression of preprotachykinin and CGRP genes in adult rat sensory neurons, *Soc. Neurosci.,* 14, 1113, 1988.

16. **Greenberg, M. E., Greene, L. A., and Ziff, E. B.,** Nerve growth factor and epidermal growth factor induce rapid transient changes in proto-oncogene transcription in PC12 cells, *J. Biol. Chem.,* 260, 14101, 1985.

17. **Tiercy, J. M. and Shooter, E. M.,** Early changes in the synthesis of nuclear and cytoplasmic proteins are induced by nerve growth factor in differentiating rat PC12 cells, *J. Cell Biol.,* 103, 2367, 1986.

18. **McGuire, J. C., Greene, L. A., and Furano, A. V.,** Nerve growth factor stimulates incorporation of fucose or glucosamine into an external glycoprotein in cultured PC12 pheochromocytoma cells, *Cell,* 15, 357, 1978.

19. **Feinstein, S. C., Dana, S. L., McConlogue, E. M., Shooter, E. M., and Coffino, P.,** Nerve growth factor rapidly induces ornithine decarboxylase mRNA in PC12 rat phaeochromocytoma cells, *Proc. Natl. Acad. Sci. U.S.A.,* 82, 5761, 1985.

20. **Lee, V. M.-Y. and Page, I. C.,** The dynamics of nerve growth factor induced microfilament and vimentin filament expression and organization in PC12 cells, *J. Neurosci.,* 4, 1705, 1984.

21. **Milbrandt, J.,** Nerve growth factor rapidly induces c-fos mRNA in PC12 rat pheochromocytoma cells, *Proc. Natl. Acad. Sci. U.S.A.,* 83, 4789, 1986.

22. **Kruijer, W., Schubert, D., and Verma, I. M.,** Induction of the proto-oncogene fos by nerve growth factor, *Proc. Natl. Acad. Sci. U.S.A.,* 82, 7330, 1985.

23. **Leonard, D. G. B., Gorham, J. D., Cole, P., Greene, L. A., and Ziff, E. B.,** A nerve growth factor-regulated mRNA encodes a new intermediate filament protein, *J. Cell Biol.,* 106, 181, 1988.

24. **Anderson, D. J. and Axel, R.,** Molecular probes for the development and plasticity of neural crest derivatives, *Cell,* 42, 649, 1985.

25. **Bandtlow, C. E., Heumann, R., Schwab, M. E., and Thoenen, H.,** Cellular localization of nerve growth factor synthesis in various organs of the peripheral nervous system, in *Neuronal Plasticity and Trophic Factors,* Biggio, G., Spano, P. F., Toffano, G., Appel, S. H., and Gessa, G. I., Eds., Fidia Research Series, Symposia in Neuroscience VII, Liviana Press, Padova, 1988.

26. **Davies, A. M., Bandtlow, C., Heumann, R., Korsching, S., Rohrer, H., and Thoenen, H.,** Timing and site of nerve growth factor synthesis in developing skin in relation to innervation and expression of the receptor, *Nature,* 326, 353, 1987.

27. **Shelton, D. L. and Reichardt, L. F.,** Expression of the β-nerve growth factor gene correlates with the density of sympathetic innervation in effector organs, *Proc. Natl. Acad. Sci. U.S.A.,* 81, 7951, 1984.

28. **Gunderson, R. W. and Barrett, J. N.,** Neuronal chemotaxis: chick dorsal root axons turn toward high concentrations of nerve growth factor, *Science,* 206, 1079, 1977.

29. **Shelton, D. L. and Reichardt, L. F.,** Studies on the regulation of β-nerve growth factor gene expression in the rat iris: the level of mRNA-encoding nerve growth factor is increased in irises placed in explant cultures *in vitro* but not in irises deprived of sensory or sympathetic innervation *in vivo, J. Cell Biol.,* 102, 1940, 1986.

30. **Rush, R. A.,** Immunohistochemical localization of endogenous nerve growth factor, *Nature,* 312, 364, 1984.

31. **Heumann, R., Korsching, S., Bandtlow, C., and Thoenen, H.,** Changes of nerve growth factor synthesis in nonneuronal cells in response to sciatic nerve transection, *J. Cell. Biol.,* 104, 1623, 1987.

32. **Heumann, R., Lindholm, D., Bandtlow, C., Meyer, M., Radeke, M. J., Misko, T. P., Shooter, E., and Thoenen, H.,** Differential regulation of mRNA encoding nerve growth factor and its receptor in rat sciatic nerve during development, degeneration and regeneration: role of macrophages, *Proc. Natl. Acad. Sci. U.S.A.,* 84, 8735, 1987.

33. **Hall, S. M.,** The effect of inhibiting Schwann cell mitosis on the reinnervation of acellular autographs in the peripheral nervous system of the mouse, *Neuropathol. Appl. Neurobiol.,* 12, 401, 1986.

34. **Smith, G. V. and Stevenson, J. A.,** Peripheral nerve grafts lacking viable Schwann cells fail to support central nervous system axonal regeneration, *Exp. Br. Res.,* 69, 299, 1988.

35. **Johnson, E. M., Jr., Taniuchi, M., and DiStefano, P. S.,** Expression and possible function of nerve growth factor receptors on Schwann cells, *TINS,* 11, 299, 1988.

36. **Buck, C. R., Martinez, H. J., Black, I. B., and Chao, M. V.,** Developmentally regulated expression of the nerve growth factor receptor gene in the periphery and brain, *Proc. Natl. Acad. Sci. U.S.A.,* 84, 3060, 1987.

37. **Yan, Q., Snider, W. D., Pinzone, J. J., and Johnson, E. M., Jr.,** Retrograde transport of nerve growth factor (NGF) in motoneurons of developing rats, *Neuron,* 1 335, 1988.

38. **Gibbs, R. B., McCabe, J. T., Buck, C. R., Chao, M. V., and Pfaff, D. W.,** Expression of NGF receptor in the rat forebrain detected with *in situ* hybridization and immunohistochemistry, submitted.

39. **Taniuchi, M., Clark, H. B., and Johnson, E. M., Jr.,** Induction of nerve growth factor receptor in Schwann cells after axotomy, *Proc. Natl. Acad. Sci. U.S.A.,* 83, 1950, 1986.

40. **DiStefano, P. J. and Johnson, E. M., Jr.,** Nerve growth factor receptors on cultured rat Schwann cells, *J. Neurocytol.,* 8, 231, 1988.

41. **Finn, P. J., Ferguson, I. A., Wilson, P. A., Vahaviolos, J., and Rush, R. A.,** Immunohistochemical evidence for the distribution of nerve growth factor in the embryonic mouse, *J. Neurocytol.,* 16, 639, 1987.

42. **Seiler, M. and Schwab, M. E.,** Specific retrograde transport of nerve growth factor (NGF) from neocortex to nucleus basalis in the rat, *Brain Res.,* 300, 33, 1984.

43. **Schwab, M. E., Otten, U., Agid, Y., and Thoenen, H.,** Nerve growth factor (NGF) in the rat CNS: absence of specific retrograde axonal transport and TH induction in locus coeruleus and substantia nigra, *Brain Res.,* 168, 473, 1979.

44. **Rennert, P. D. and Heinrich, G.,** Nerve growth factor mRNA in brain: localization by *in situ* hybridization, *Biochem. Biophys. Res. Commun.,* 138, 813, 1986.

45. **Ayer-LeLievre, C., Olson, L., Ebendal, T., Seiger, A., and Persson, H.,** Expression of the β-nerve growth factor gene in hippocampal neurons, *Science,* 240, 1339, 1988.

46. **Friedman, P. L., Larhammar, D., Holets, V. R., Gonzalez-Carvajal, M., Yu, Z. Y., Persson, H., and Wittemore, S. R.,** Rat β-NGF sequence and sites of synthesis in adult CNS, *Soc. Neurosci.,* 14, 827, 1988.

47. **Hayes, R. C., Rosenberg, M. B., Higgins, G. A., Chen, K. S., Gage, F. H., and Armstrong, D. M.,** *In situ* hybridization of nerve growth factor mRNA in adult rat brain, *Soc. Neurosci.,* 14, 684, 1988.

48. **Lindsay, R. M.,** Adult rat brain astrocytes support survival of both NGF-dependent and NGF insensitive neurones, *Nature,* 282, 80, 1979.

49. **Furukawa, S., Furukawa, Y., Satoyoshi, E., and Hayashi, K.,** Synthesis/secretion of nerve growth factor is associated with cell growth in cultured mouse astroglial cells, *Biochem. Biophys. Res. Commun.,* 142, 395, 1987.

50. **Korsching, S., Auberger, G., Heumann, R., Scott, J., and Thoenen, H.,** Levels of nerve growth factor and its mRNA on the central nervous system of the rat correlate with cholinergic innervation, *EMBO J.,* 4, 1389, 1985.

51. **Korsching, S.,** The role of nerve growth factor in the central nervous system, *TINS,* 570, 1986.

52. **Lorez, H. P., von Frankenberg, M., Weskamp, G., and Otten, U.,** Effect of bilateral decortication on nerve growth factor content in basal nucleus and neostriatum of adult rat brain, *Brain Res.,* 454, 355, 1988.

53. **Gage, F. H., Olejniczak, P., and Armstrong, D. M.,** Astrocytes are important for sprouting in the septohippocampal circuit, *Exp. Neurol.,* 102, 22, 1988.

54. **Schatteman, G. C., Gibbs, L., Lanahan, A. A., Claude, P., and Bothwell, M.,** Expression of nerve growth factor receptor in developing and adult primate central nervous system, *J. Neurosci.,* 8, 860, 1988.

55. **Whittemore, S. R., Ebendal, T., Larkfors, L., Olson, L., Seiger, A., Stromberg, I., and Persson, H.,** Developmental and regional expression of β-nerve growth factor messenger RNA and protein in rat central nervous system, *Proc. Natl. Acad. Sci. U.S.A.,* 83, 817, 1986.

56. **Taniuchi, M., Clarke, H. B., Schweitzer, J. B., and Johnson, E. M., Jr.,** Expression of nerve growth factor receptors by Schwann cells of axotomized peripheral nerves: ultrastructural location, suppression by axonal contact and binding properties, *J. Neurosci.,* 8, 664, 1988.

57. **Cavicchiolio, L., Vantini, G., Flanigan, T., Walsh, F., Fusco, M., Bigon, E., Benvegnu, D., and Leon, A.,** NGF enhances the expression of NGF receptor mRNA *in vivo, Soc. Neurosci.,* 14, 1248, 1988.

58. **Hefti, F.,** Nerve growth factor promotes survival of septal cholinergic neurons after fimbrial transection, *J. Neurosci.,* 6, 2155, 1986.

59. **Kromer, L. F.,** Nerve growth factor treatment after brain injury prevents neuronal death, *Science,* 235, 214, 1987.

60. **Hefti, F., Dravid, A., and Hartikka, J.,** Chronic intraventricular injections of nerve growth factor elevate hippocampal choline acetyltransferase activity in adult rats with partial septohippocampal lesions, *Brain Res.,* 293, 305, 1984.

61. **Hefti, F., Hartikka, J., Eckenstein, F., Gnahn, H., Heumann, R., and Schwab, M.,** NGF increases choline acetyltransferase but not survival or fiber outgrowth of cultured fetal septal cholinergic neurons, *Neuroscience,* 14, 55, 1985.

62. **Ishii, D. N.,** Insulin like growth factor II gene expression in muscle: relationship to synapse elimination and nerve regeneration, *Soc. Neurosci.,* 13, 1211, 1987.

63. **Recio-Pinto, E. and Ishii, D. N.,** Insulin and related growth factors: effects on the nervous system and mechanism for neurite growth and regeneration, *Neurochem. Int.,* 12, 397, 1988.

64. **Recio-Pinto, E., Rechler, M. M., and Ishii, D. N.,** Effects of insulin, insulin-like growth factor-II and nerve growth factor on neurite formation and survival in cultured sympathetic and sensory neurons, *J. Neurosci.,* 6, 1211, 1986.

65. **Mill, J. F., Chao, M. V., and Ishii, D. N.,** Insulin, insulin-like growth factor II and nerve growth factor effects on tubulin mRNA levels and neurite formation, *Proc. Natl. Acad. Sci. U.S.A.,* 82, 7126, 1985.

66. **Recio-Pinto, E., Land, F. F., and Ishii, D. N.,** Insulin and insulin-like growth factor binding and the neurite formation response in cultured human neuroblastoma cells. *Proc. Natl. Acad. Sci. U.S.A.,* 81, 2562, 1984.

67. **Wang, C., Wible, B., Angelides, K., and Ishii, D.,** Insulin and insulin like growth factor I increase neurofilament mRNA levels and neurite formation, *Soc. Neurosci.,* 14, 1169, 1988.

68. **Glazer, G. W. and Ishii, D. N.,** Insulin, insulin-like growth factor I and nerve growth factor stimulate neurite formation in rat spinal cultures, *Soc. Neurosci.,* 14, 1040, 1988.

69. **Lund, P. K., Moats-Staats, B. M., Hynes, M. A., Simmons, J. G., Jansen, M., D'Ercole, A. J., and van Wyck, J. J.,** Somatomedin-C/insulin-like growth factor-I and insulin growth factor-II mRNAs in rat fetal and adult tissues, *J. Biol. Chem.,* 261, 14539, 1986.

70. **Ballotti, R., Nielsen, F. C., Pringle, N., Kowalski, A., Richardson, W. D., von Obberghen, A., and Gammeltoft, S.,** Insulin like growth factor I in cultured rat astrocytes: expression of the gene and receptor tyrosine kinase, *EMBO J.,* 6, 3533, 1987.

71. **Rosen, K. M. and Villa-Komaroff, L.,** Distribution of insulin-like growth factor II mRNA in the central nervous system, *Soc. Neurosci.,* 13, 1706, 1987.

72. **Rotwein, P., Burgess, S. K., Milbrandt, J. D., and Krause, J. E.,** Differential expression of insulin-like growth factor genes in rat central nervous system, *Proc. Natl. Acad. Sci. U.S.A.,* 85, 265, 1988.

73. **Baskin, D. G., Wilcox, B. J., Figlewicz, D. P., and Dorsa, D. M.,** Insulin and insulin like growth factors in the CNS, *TINS,* 11, 107, 1988.

74. **DiCicco-Bloom, E. and Black, I. B.,** Insulin growth factors regulate the mitotic cycle in cultured rat sympathetic neuroblasts, *Proc. Natl. Acad. Sci., U.S.A.,* 85, 4066, 1988.

75. **Dubois-Dalcq, M.,** Characterization of a slowly proliferative cell along the oligodendrocyte differentiation pathway, *EMBO J.,* 6, 2587, 1987.

76. **Toran-Allerand, C. D., Anderson, J. P., and Bentham, W.,** Insulin influences astroglial morphology and GFAP expression *in vitro, Soc. Neurosci.,* 14, 1247, 1988.

77. **Young, W. S., III,** Periventricular hypothalamic cells in the rat brain contain insulin mRNA, *Neuropeptides,* 8, 93, 1986.

78. **Hansson, H. A., Rozell, B., and Skottner, A.,** Rapid axoplasmic transport of insulin-like growth factor I in sciatic nerve of adult rat, *Cell Tissue Res.,* 247, 241, 1987.

79. **Irminger, J.-C., Rosen, K. M., Humbel, R. E., and Villa-Komaroff, L.,** Tissue-specific expression of insulin-like growth factor II mRNAs with distinct 5' untranslated regions, *Proc. Natl. Acad. Sci. U.S.A.,* 84, 6330, 1987.

80. **Mudd, L. M., Masters, B. A., and Raizada, M. K.,** Insulin and related growth factors: effects on the nervous system and mechanism for neurite outgrowth and regeneration, *Neurochem. Int.,* 12, 415, 1988.

81. **Unger, J., McNeill, T. H., Moxley, R. T., III, and Livingston, J. N.,** Insulin receptors and neuropeptides in limbic and hypothalamic areas of the rat brain, *Soc. Neurosci.,* 14, 103, 1988.

82. **Carpenter, G. J. and Cohen, S.,** Epidermal growth factor, *Ann. Rev. Biochem.,* 48, 193, 1979.

83. **Almazar, G., Honegger, P., Metthiey, J.-M., and Guentert-Lauber, B.,** Epidermal growth factor and bovine growth hormone stimulate differentiation and myelination of brain cell aggregates in culture, *Dev. Brain Res.,* 21, 251, 1985.

84. **Ran, W., Dean, M., Levine, R. A., Henkle, C., and Campisi, J.,** Induction of c-fos and c-myc mRNA by epidermal growth factor or calcium ionophore is cAMP dependent, *Proc. Natl. Acad. Sci. U.S.A,* 83, 8216, 1986.

85. **Morrison, R. S., Kornblum, H. I., Leslie, F. M., and Bradshaw, R. A.,** Trophic stimulation of cultured neurons from neonatal rat brain by epidermal growth factor, *Science,* 238, 72, 1987.

86. **Rall, L. B., Scott, J., and Bell, I.,** Mouse prepro-epidermal growth factor synthesis by the kidney and other tissues, *Nature,* 313, 228, 1985.

87. **Lazar, L. M., Roberts, J. L., and Blum, M.,** Regional distribution of epidermal growth factor mRNA in the mammalian central nervous system, *Soc. Neurosci.,* 14, 1162, 1988.

88. **Fallon, J. H., Seroogy, K. B., Loughlin, S. E., Morrison, R. S., Bradshaw, R. A., Knauer, D. J., and Cunningham, D. D.,** Epidermal growth factor immunoreactive material in the CNS: location and development, *Science,* 224, 1107, 1984.

89. **Marquardt, H., Hunkapiller, M. W., Hood, L. E., and Todaro, G. J.,** Rat transforming growth factor type I: structure and relation to epidermal growth factor, *Science,* 223, 1079, 1984.

90. **Massague, J.,** The epidermal growth factor like transforming growth factors, *J. Biol. Chem.,* 258, 12614, 1983.

91. **Wilcox, J. N. and Derynck, R.,** Localization of cells synthesizing transforming growth factor-alpha mRNA in the mouse brain, *J. Neurosci.,* 8, 1901, 1988.

92. **Weeks, D. L. and Melton, D. A.,** A maternal mRNA localized to the vegetal hemisphere in Xenopus eggs codes for a growth factor related to TGF-β, *Cell,* 51, 861, 1987.

93. **Gomez-Penilla, F., Knauer, D. J., and Nieto-Sampedro, M.,** Epidermal growth factor receptor immunoreactivity in rat brain: development and cellular localization, *Brain Res.,* 438, 385, 1988.

94. **Nieto-Sampedro, M., Gomez-Pinella, F., Knauer, D. J., and Broderick, J. T.,** Epidermal growth factor receptor in rat brain astrocytes. Response to injury, *Neurosci. Lett.,* 91, 276, 1988.

95. **Nieto-Sampedro, M.,** Astrocyte mitogen inhibitor related to epidermal growth factor receptor, *Science,* 240, 1784, 1988.

96. **Libermann, T. A., Nusbaum, H. R., Razon, N., Kris, R., Lax, I., Soreq, H., Whittle, N., Waterfield, M. D., Ullrich, A., and Schlessinger, J.,** Amplification, enhanced expression and possible rearrangement of EGF receptor gene in primary human brain tumours of glial origin, *Nature,* 313, 144, 1985.

97. **Downward, J., Yarden, Y., Mayes, E., Scrace, G., Totty, N., Stockwell, P., Ullrich, A., Schlessinger, J., and Waterfield, M. D.,** Close similarity of epidermal growth factor receptor and v-erb-B oncogene protein sequences, *Nature,* 307, 521, 1984.

98. **Foote, A. M., Rost, N., Chauvin, C., Nissou, M. F., and Benabid, A. L.,** Distribution of proto-oncogene expression in rat brain, *Soc. Neurosci.,* 13, 1706, 1987.

99. **Bargmann, C. I., Hung, M.-C., and Weinberg, R. A.,** The neu oncogene encodes an epidermal growth factor receptor related protein, *Nature,* 319, 226, 1986.

100. **Kokai, Y., Cohen, J. A., Drebin, J. A., and Greene, M. I.,** Stage and tissue specific expression of the neu oncogene in rat development, *Proc. Natl. Acad. Sci. U.S.A.,* 84, 8498, 1987.

101. **Pruss, R. M., Bartlett, P. F., Gavrilovic, J., Lisak, R. P., and Rattray, S.,** Mitogens for glial cells: a comparison of the response of cultured astrocytes, oligodendrocytes and Schwann cells, *Dev. Brain Res.,* 2, 19, 1982.

102. **Kniss, D. A. and Burry, R. W.,** Serum and fibroblast growth factor stimulate quiescent astrocytes to re-enter the cell cycle, *Brain Res.,* 439, 281, 1988.

103. **Walicke, P. A.,** Basic and acidic fibroblast growth factors have trophic effects on neurons from multiple CNS regions, *J. Neurosci.,* 8, 2618, 1988.

104. **Walicke, P. A. and Baird, A.,** Neurotrophic effects of basic and acidic fibroblast growth factors are not mediated through glial cells, *Dev. Brain Res.,* 40, 71, 1988.

105. **Walicke, P. A. and Baird, A.,** Trophic effects of fibroblast growth factor on neural tissue, *Progress in Brain Research,* 1988, in press.

106. **Eccleston, P. A. and Silberberg, D. H.,** Fibroblast growth factor is a mitogen for oligodendrocytes *in vitro, Dev. Brain Res.,* 21, 315, 1985.

107. **Togari, A., Dickins, G., Kuzuya, H., and Guroff, G.,** The effect of fibroblast growth factor on PC12 cells, *J. Neurosci.,* 5, 307, 1985.

108. **Finklestein, S. P., Apostolides, P. J., Caday, C. G., Prosser, J., Philips, M. F., and Klagsbrun, M.,** Increased basic fibroblast growth factor (BFGF) immunoreactivity at the site of focal brain wounds, *Brain Res.,* 460, 253, 1988.

109. **Unsicker, K., Krisch, B., Otten, J., and Thoenen, H.,** Nerve growth factor induced fiber outgrowth from isolated rat adrenal chromaffin cells, *Proc. Natl. Acad. Sci. U.S.A.,* 75, 3498, 1978.

110. **Janet, T., Miehe, M., Pettmann, B., Labourdette, G., and Sensenbrenner, M.,** Ultrastructural localization of fibroblast growth factor in neurons of rat brain, *Neurosci. Lett.,* 80, 153, 1987.

111. **Needles, D. L. and Cotman, C. W.,** Basic fibroblast growth factor (with heparin) increases the survival of rat dentate granule cells in culture, *Soc. Neurosci.,* 14, 363, 1988.

112. **Caday, C. G., Mirzabegian, A., Prosser, J., Klagsbrun, M., and Finklestein, S. P.,** Fibroblast growth factor (FGF) levels in the developing brain, *Soc. Neurosci.,* 14, 363, 1988.

113. **Herblin, W. F., Krause, R. G., and Schwaber, J. S.,** Distribution of binding sites for basic fibroblast growth factor in rat brain, *Soc. Neurosci.,* 14, 104, 1988.

114. **Huang, S. S., Tsai, C. C., Adams, S. P., and Huang, J. S.,** Neuron localization and neuroblastoma cell expression of brain derived growth factor, *Biochem. Biophys. Res. Commun.,* 144, 81, 1987.

115. **Senior, R. M., Huang, S. S., Griffin, G. L., and Huang, J. S.,** Brain derived growth factor is a chemoattractant for fibroblasts and astroglial cells, *Biochem. Biophys. Res. Commun.,* 141, 67, 1986.

116. **Jakobovits, A., Shackleford, G. M., Varmus, H. E., and Martin, G. R.,** Two proto-oncogenes implicated in mammary carcinogenesis, int-1 and int-2, are independently regulated during mouse development, *Proc. Natl. Acad. Sci. U.S.A.,* 83, 7806, 1986.

117. **Smith, R., Peters, G., and Dickson, C.,** Multiple RNAs expressed from the int-2 gene in mouse embryonal carcinoma cell lines encode a protein with homology to fibroblast growth factor, *EMBO J.,* 7, 1013, 1988.

118. **Dickson, C. and Peters, G.,** Potential oncogene product related to growth factors, *Nature,* 326, 833, 1987.

119. **Wilkinson, D. G., Peters, G., Dickson, C., and McMahon, A. P.,** Expression of the FGF-related proto-oncogene int-1 during gastrulation and neurulation in the mouse, *EMBO J.,* 7, 691, 1988.

120. **Wilkinson, D. G., Bailes, J. A., and McMahon, A. P.,** Expression of the proto-oncogene int-2 is restricted to specific neural cells in the developing mouse embryo, *Cell,* 50, 79, 1987.

121. **Betsholtz, C., Johnsson, A., Heldin, C.-H., Westermark, B., Lund, P., Urden, M. S., Eddy, R., Shows, T. B., Philpott K., Mellor, A., Knott, T. J., and Scott, J.,** cDNA sequence and chromosomal localization of human platelet derived growth factor A-chain and its expression in tumor cell lines, *Nature,* 320, 695, 1986.

122. **Richardson, W. D., Pringle, N., Mosley, M. J., Westermark, B., and Dubois-Dalcq, M.,** A role for platelet-derived growth factor in normal gliogenesis in the CNS, *Cell,* 53, 309, 1988.

123. **Pringle, N., Collarini, E. J., Mosley, M. J., Heldin, C.-H., Westermark, B., and Richardson, W. D.,** PDGF A chain homodimers drive proliferation of bipotential O2A glial progenitor cells in developing rat optic nerve, in press.

124. **Waterfield, M. D., Scrace, G. T., Whittle, N., Stroobant, P., Johnsson, A., Wasteson, A., Westermark, B., Heldin, C.-H., Huang, J. S., and Deuel, T. F.,** Platelet derived growth factor is structurally related to the putative transforming protein p28 sis of simian sarcoma virus, *Nature,* 304, 35, 1983.

125. **Doolittle, R. F., Hunkapillier, M. W., Hook, L. E., Deva, S. G., Robbins, K. C., Aaronson, S. A., and Antoniades, H. N.,** Simian sarcoma virus onc, v-sis, is derived from the gene (or genes) encoding a platelet derived growth factor, *Science,* 221, 275, 1983.

126. **Ross, R., Raines, E. W., and Bowen-Pope, D. F.,** The biology of platelet-derived growth factor, *Cell,* 46, 155, 1986.

127. **Alema, S., Casalbore, P., Agostini, E., and Tato, F.,** Differentiation of PC12 phaeochromocytoma cells induced by v-src oncogene, *Nature,* 316, 557, 1985.

128. **Aubry, M. and Maness, P. F.,** Developmental regulation of protein tyrosine phosphorylation in rat brain, submitted.

129. **Cartwright, C. A., Simantov, R., Kaplan, P. L., Hunter, T., and Eckhart, W.,** Alterations in pp60c-src accompany differentiation of neurons from rat embryo striatum, *Mol. Cell Biol.,* 7, 1830, 1987.

130. **Fults, D. W., Towle, A. C., Lauder, J. M., and Maness, P. F.,** pp60c-src in ther developing cerebellum, *Mol. Cell Biol.,* 5, 27, 1985.

131. **Sorge, L. K., Levy, B. T., and Maness, P. F.,** pp60 c-src is developmentally regulated in the neural retina, *Cell,* 36, 249, 1984.

132. **Maness, P. F.,** pp60c-src encoded by the proto-oncogene c-src is a product of sensory neurons, *J. Neurosci. Res.,* 16, 127, 1986.

133. **Maness, P. F., Aubry, M., Shores, C. G., Frame, L., and Pfenninger, K. H.,** c-src gene products in developing rat brain is enriched in nerve growth cone membranes, *Proc. Natl. Acad. Sci. U.S.A.,* in press.

134. **Cotton, P. C. and Brugge, J. S.,** Neural tissues express high levels of the cellular src gene product pp60 c-src, *Mol. Cell Biol.,* 3, 1157, 1983.

135. **Walaas, S. I., Lustig, A., Greengard, P., and Brugge, J. S.,** Widespread distribution of the c-src gene product in nerve cells and axon terminals in the adult rat brain, *Mol. Brain Res.,* 3, 215, 1988.

136. **Lasher, R. S., Erickson, P. F., Mena, E. E., and Cotman, C. W.,** The binding of a monoclonal antibody reactive with pp60 v-src to the rat CNS both *in vitro* and *in vivo, Brain Res.,* 452, 184, 1988.

137. **Brugge, J. S., Cotton, P. C., Queral, A. E., Barrett, J. N., Nonner, D., and Keane, R. W.,** Neurones express high levels of a structurally modified activated form of pp60 c-src, *Nature,* 316, 554, 1985.

138. **Ingraham, C. A. and Maness, P. F.,** Expression of c-src+ mRNA by ganglion and amacrine cells of chick neural retina revealed by *in situ* hybridization, *Soc. Neurosci.,* 14, 635, 1988.

139. **Ross, C. A., Pearson, R. C. A., Wright, G., and Snyder, S. H.,** Expression of brain specific src(+) oncogene in rat brain, studied with *in situ* hybridization, *Soc. Neurosci.,* 14, 753, 1988.

140. **Ross, C. A., Wright, G. E., Resh, M. D., Pearson, R. C. A., and Snyder, S. H.,** Brain-specific src oncogene mRNA mapped in rat brain by *in situ* hybridization, *Proc. Natl. Acad. Sci. U.S.A.,* in press.

141. **Sap, J., Munoz, A., Damm, K., Goldberg, Y., Ghysdael, J., Leutz, A., Beug, H., and Vennstrom, B.,** The c-erb-A protein is a high affinity receptor for thyroid hormone, *Nature,* 324, 635, 1986.

142. **de Vos, A. M., Tong, L., Milburn, M. V., Matias, P. M., Jancarik, J., Noguchi, S., Nishimura, S., Miura, K., Ohtsuka, E., and Kim, S.-H.,** Three-dimensional structure of an oncogene protein: catalytic domain of human c-H-ras p21, *Science,* 239, 888, 1988.

143. **Dolphin, A. C.,** Is p21-ras a real G protein? *TINS,* 11, 287, 1988.

144. **Guerrero, I., Pellicer, A., and Burstein, D. E.,** Dissociation of c-fos from ODC expression and neuronal differentiation in a PC12 subline stably transfected with an inducible n-ras oncogene, *Biochem. Biophys. Res. Commun.,* 150, 1185, 1988.

145. **Noda, M., Ko, M., Ogura, A., Liu, D. G., Amano, T., Takano, T., and Ikawa, Y.,** Sarcoma viruses carrying ras oncogenes induce differentiation associated properties in a neuronal cell line, *Nature,* 318, 73, 1985.

146. **Bar-Sagi, D. and Feramisco, J. R.,** Microinjection of the ras oncogene protein into PC12 cells induces morphological differentiation, *Cell,* 42, 841, 1985.

147. **Cole, M. D.,** The myc oncogene: its role in transformation and differentiation, *Ann. Rev. Genet.,* 20, 361, 1986.

148. **Maruyama, K., Shiavi, S. C., Huse, W., Johnson, G. L., and Ruley, H. E.,** myc and E1A oncogenes alter responses of PC12 cells to nerve growth factor and block differentiation, *Oncogene,* 1, 361, 1987.

149. **Brakefield, X. O. and Stern, D. S.,** Oncogenes in neural tumors, *TINS,* 9, 1986.

150. **Squire, J., Goddard, A. D., Canton, M., Becker, A., Phillips, R. A., and Gallie, B. L.,** Tumor induction by the retinoblastoma mutation is independent of n-myc expression, *Nature,* 322, 555, 1986.

151. **Zimmerman, K. A., Yancopoulos, G. D., Collum, R. G., Smith, R. K., Kohl, N. E., Denis, K. A., Nau, M. M., Witte, O. N., Toran-Allerand, D., Gee, C. E., Minna, J. D., and Alt, F. W.,** Differential expression of myc family genes during murine development, *Nature,* 319, 780, 1986.

152. **Kasik, J. W., Wan, Y.-J. Y., and Ozato, K.,** A burst of c-fos gene expression in the mouse occurs at birth, *Mol. Cell. Biol.,* 7, 3349, 1987.

153. **Stein-Izsak, C., Cohen, I., Chesselet, M.-F., Murray, M., and Schwartz, M.,** Expression of proto-oncogenes fos and myc increases after injury to the fish optic nerve: *in vitro* and *in situ* studies, *Soc. Neurosci.,* 13, 195, 1987.

154. **Curran, T. and Morgan, J. I.,** Memories of fos, *BioEssays,* 7, 255, 1987.

155. **Ruppert, C., Goldowitz, D., and Wille, W.,** Proto-oncogene c-myc is expressed in cerebellar neurons at different developmental stages, *EMBO J.,* 5, 1897, 1986.

156. **Sambucetti, L. C. and Curran, T.,** The fos protein complex is associated with DNA in isolated nuclei and binds to DNA cellulose, *Science,* 234, 1417, 1986.

157. **Franza, B. R., Jr., Rauscher, F. J., III, Josephs, S. F., and Curran, T.,** The fos complex and fos-related antigens recognize sequence elements that contain AP-1 binding sites, *Science,* 239, 1150, 1988.

158. **Rauscher, F. J., III, Sambucetti, L. C., Curran, T., Distel, R. J., and Spiegelman, B. M.,** Common DNA binding site for fos protein complexes and transcription factor AP-1, *Cell,* 52, 471, 1988.

159. **Verma, I. M.,** Proto-oncogene fos: a multifaceted gene, *TIG,* 93, 1986.

160. **Goelet, P., Castellucci, V. F., Schacher, S., and Kandel, E. R.,** The long and the short of long-term memory—a molecular framework, *Nature,* 322, 419, 1986.

161. **Dragunow, M. and Robertson, H. A.,** Kindling stimulation induces c-fos protein(s) in granule cells of the rat dentate gyrus, *Nature,* 329, 441, 1987.

162. **Dragunow, M. and Robertson, H. A.,** Localization and induction of c-fos protein-like immunoreactive material in the nuclei of adult mammalian neurons, *Brain Res.,* 440, 252, 1988.

163. **Dragunow, M. and Robertson, H. A.,** Brain injury induces c-fos protein(s) in nerve and glial-like cells in adult mammalian brain, *Brain Res.,* 455, 295, 1988.

164. **Hunt, S. P., Pini, A., and Evan, G.,** Induction of c-fos-like protein in spinal cord neurons following sensory stimulation, *Nature,* 328, 632, 1987.

165. **Morgan, J. I., Cohen, D. R., Hempstead, J. L., and Curran, T.,** Mapping patterns of c-fos expression in the central nervous system after seizure, *Science,* 237, 192, 1987.

166. **Gubits, R. M., Hazelton, J. L., and Simantov, R.,** Variations in c-fos gene expression during rat brain development, *Mol. Brain Res.,* 3, 197, 1988.

167. **Cohen, D. R., Morgan, J. I., Hempstead, J. L., and Curran, T.,** Rapid induction of proto-oncogene expression in the CNS, in *Mechanisms of Control of Gene Expression,* Alan R. Liss, New York, 1988, 327.

168. **Sequier, J. M., Malherbe, P., Hunziker, W., Mohler, H., and Richards, J. G.,** *In situ* hybridization histochemistry as a tool to study the regional expression of gaba/benzodiazep-ine receptors, the proto-oncogene c-fos and calbindin d28 in tissue sections, Conference on Molecular Neurobiology, NIMH Neurosciences Research Branch, Bethesda, MD, 1988.

169. **Gall, C.,** Seizures induce dramatic and distinctly different changes in enkephalin, dynorphin, and cholecystokinin immunoreactivities in mouse hippocampal mossy fibers, *J. Neurosci.,* 8, 1852, 1988.

170. **Gall, C., Arai, A., and White, J.,** Localization of increased c-fos mRNA content in rat CNS following recurrent seizures, *Soc. Neurosci.,* 14, 1161, 1988.

171. **White, J. D. and Gall, C. M.,** Differential regulation of neuropeptide and proto-oncogene mRNA content in the hippocampus following recurrent seizures, *Mol. Brain Res.,* 3, 21, 1987.

172. **Chang, S. L., Harlan, R. E., and Squinto, S. P.,** Morphine increases c-fos mRNA in rat caudate-putamen, *Soc. Neurosci.,* 14, 752, 1988.

173. **Sagar, S. M., Sharp, F. R., and Curran, T.,** Expression of c-fos protein in brain: metabolic mapping at the cellular level, *Science,* 240, 1228, 1988.

174. **Young, W. S., III and Zoeller, R. T.,** Neuroendocrine gene expression in the hypothalamus, *Cell. Mol. Neurobiol.,* 7, 353, 1987.

175. **Sherman, T. G., Akil, H., and Watson, S. J.,** Vasopressin mRNA expression: a northern and *in situ* hybridization analysis, in *Vasopressin,* Schrier, R. W., Ed., Raven Press, New York, 1985, 475.

176. **Gall, C. M., Pico, C. M., and Lauterborn, J. C.,** Focal hippocampal lesions induce seizures and long lasting changes in mossy fiber enkephalin and CCK immunoreactivity, *Peptides,* 9, 79, 1988.

177. **Jorgensen, M. B., Deckert, J., Wright, D. C., and Gehlert, D. R.,** Delayed C-fos proto-oncogene expression in the rat hippocampus induced by transient global cerebral ischemia: an *in situ* hibridization study, *Brain Res.,* in press.

Chapter 6

THE USE OF *IN SITU* TRANSCRIPTION IN THE STUDY OF GENE EXPRESSION

James Eberwine, Ines Zangger, Russell Van Gelder, Chris Evans, and Laurence Tecott

TABLE OF CONTENTS

I. INTRODUCTION

Changes in gene expression result in an alteration in cellular function manifested by changes in cellular structure, action potential, secretion of peptide and/or nonpeptide neuromodulators and neurotransmitters, and a host of other physiologic processes (Figure 1). Such changes in cellular activity, better termed cellular responsiveness, determine a cascade of events which alter tissue and organismal functioning. To understand these resultant multicellular events, it is important to understand how the single cell is regulated by changes in the concentration of specific modulators. In particular, the ability to localize the site of expression of particular genes provides insight into what factors may modulate the amount of gene product and how these modulators function.

The use of *in situ* hybridization histochemistry provides a means by which gene expression can be localized to specific tissues,[1,2] specific cells within these tissues,[3-5] and ultimately to specific subcellular sites within individual cells.[6] The limitations of *in situ* hybridization, in particular the intensity of signal, make it difficult to visualize mRNAs within a short period of time. Indeed for some mRNAs it can take weeks to several months to determine the cellular localization of the mRNA within a heterogeneous cell population.[7] These signals are determined by the abundance of the mRNA being examined, the type of probe (DNA or RNA), its specific activity, and its hybridization efficiency (length, G-C content, etc.).

In an effort to allay this concern about intensity of the *in situ* hybridization signal that is generated we have developed a technique called *in situ* transcription. This technique requires the initial hybridization of a primer (usually an oligonucleotide) to the tissue section followed by the synthesis of cDNA, *in situ*, using avian myeloblastosis virus reverse transcriptase[8,9] and deoxyribonucleotide triphosphates[10] (Figure 2). The cDNA can be detected either by incorporating radioactively labeled deoxyribonucleotide triphosphates into the growing cDNA chain followed by autoradiography or by incorporating chemically modified bases followed by enzymatic visualization (e.g., biotin-labeled deoxynucleotide triphosphates with subsequent detection using alkaline phosphatase coupled to avidin). This chapter reviews the optimization of the *in situ* transcription protocol, the use of *in situ* transcription to localize mRNA molecules in tissue sections, the electrophoretic analysis of the *in situ* transcription-derived cDNA transcripts, and the use of *in situ* transcription for the nonradioactive localization of mRNA molecules. These topics will be illustrated with results generated from use of two probes used in the establishment of the technique: an oligonucleotide directed against proopiomelanocortin (POMC) mRNA,[11,12] which is present in the intermediate and anterior lobes of the rat pituitary, as well as with an oligonucleotide directed against vasopressin mRNA[13] which is present in the hypothalamus of the mouse brain.

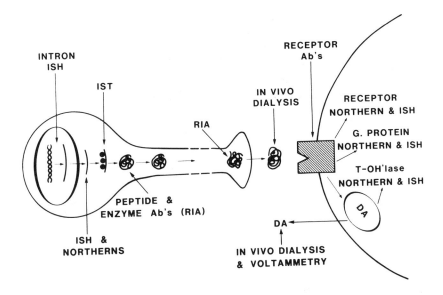

FIGURE 1. A hypothetical peptide secreting neuron is depicted that is impinging upon a dopamine synthesizing neuron where regulation can occur and how such regulation can be measured. *In situ* hybridization is abbreviated ISH. Wherever ISH is shown, *in situ* transcription (IST) can also be used. The one place in this scheme of gene expression where IST may supply unique information is in measuring the "translatability" of mRNA, discussed later.

II. OPTIMIZATION OF THE *IN SITU* TRANSCRIPTION PROTOCOL FOR USE IN mRNA LOCALIZATION STUDIES

Optimization of the *in situ* transcription protocol requires optimization of hybridization of the specific primer to the tissue section, of the cDNA synthesis reaction, and of the washing protocol to reduce the background resulting from the *in situ* transcription reaction.

A. Optimization of Specific Hybridization

In situ transcription requires the existence of a primer template complex for the reverse transcriptase to initiate cDNA synthesis. This complex is generated by hybridizing a specific primer to the tissue section of interest. This step is essentially the same as that of *in situ* hybridization. As is the case for *in situ* hybridization either a DNA[14] or RNA[15] molecule can serve as primer. Some factors that must be taken into account when optimizing this parameter of the *in situ* transcription procedure include: tissue preparation, permeabilization of the tissue, length of probe, hybridization temperature, salt and formamide concentration in the hybridization buffer, and hybridization time. In general if the tissue is fresh-frozen, cryostat cut into 11 μm sections and post-fixed with 3% paraformaldehyde for 5 min, there is little need to worry about permeabilization

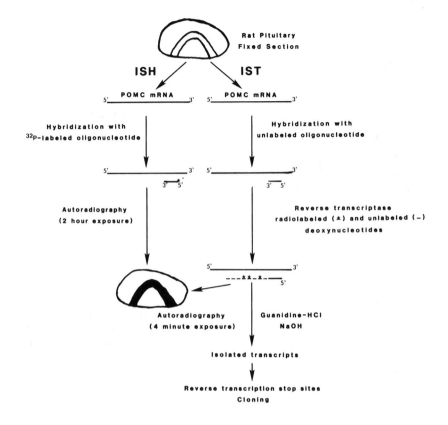

FIGURE 2. Illustration of the similarities and differences between *in situ* hybridization and *in situ* transcription.

procedures for the tissue since there seems to be little problem using the specimen for *in situ* hybridization or *in situ* transcription.

Our experience, as has been the case with other investigators,[16] has demonstrated good results using oligonucleotides which range in size from 25 to 53 bases in length. Other length oligonucleotides should also prove satisfactory, keeping the following caveats in mind: (1) If the oligonucleotide is too short it may not stay hybridized during the reverse transcription reaction (when the salt concentration is reduced), thus reducing the resultant *in situ* transcription signal and (2) if the oligonucleotide is too long it may span a region of the mRNA molecule that is associated with a cellular component such as a ribosome, etc., thus making it difficult for the 3'-end of the oligonucleotide to form a stable hybrid thereby inhibiting the initiation of reverse transcription. To optimize the other factors it is best to perform empirical determinations by setting up appropriate time course and concentration curve experiments.

The amount of primer necessary to saturate the mRNA within a section during a given time of hybridization must be determined empirically for every oligonu-cleotide and is probably dependent upon the composition of the hybridization buffer. Equations have been developed to provide an estimate of the amount of oligonucleotide needed to saturate the mRNA tethered on a solid support during a given time period, but those calculations have generally been determined from an examination of solution hybridization[17-19] parameters rather than solid phase[20] hybridizations. While these parameters are similar they are not identical, hence the need to empirically determine the amount of primer. Using the hybridization parameters (660 mM NaCl, 50% formamide, 10 mM Tris pH 7.4, 1× Denhardts, 1 mM EDTA, 0.01% tRNA, 0.05% sodium pyrophosphate, hybridize at 37°C for 24 h), a probe concentration 250,000 times larger then the amount of anticipated mRNA within the section is sufficient to provide a good *in situ* hybridization and *in situ* transcription signal.[21] This number was derived by empirically determin-ing that a primer concentration of 2.5 ng/μl gave an optimal *in situ* transcription signal, and that in an 11-μm thick section containing the intermediate and anterior lobe of the rat pituitary there should be approximately 1 pg of POMC mRNA. This calculation does not take into account the likely problems of diffusion of primer and enzyme into the tissue section.

After the hybridization step, the remaining unhybridized primer as well as the hybridization buffer must be removed from the section before the reverse transcription reaction can be performed. This is accomplished by washing the sections in 2× SSC for 1 h at room temperature, 0.5× SSC for 5 to 6 h at 40°C, and 1× *in situ* transcription buffer (50 mM Tris-HCl, pH 8.3, 6 mM MgCl$_2$; and 120 mM KCl) for 10 min at room temperature. The *in situ* transcription buffer is removed by aspiration and replaced with the buffer containing all of the components necessary to perform cDNA synthesis as described in the following section.

B. Optimization of the Reverse Transcription Reaction

Optimization of this aspect of the *in situ* transcription reaction is critically important to the generation of the *in situ* transcription signal. The difficulty inherent in using the manufacturer's suggested conditions for this reaction derives from the fact that the reverse transcriptasein the *in situ* transcription is essentially a solid-phase reaction rather than the usual solution phase reaction (Figure 3). The mechanics by which this alters enzyme activity are unclear; yet, by changing the concentration of various components of the reaction buffer, a significantly increased signal can be generated.

The *in situ* transcription reaction buffer that we use contains:

$$50 \text{ m}M \quad \text{Tris-HCl (pH 8.3)}$$
$$6 \text{ m}M \quad \text{MgCl}_2$$
$$120 \text{ m}M \quad \text{KCl}$$
$$7 \text{ m}M \quad \text{Dithiothreitol}$$
$$250 \text{ }M \quad \text{dATP, dGTP, TTP}$$

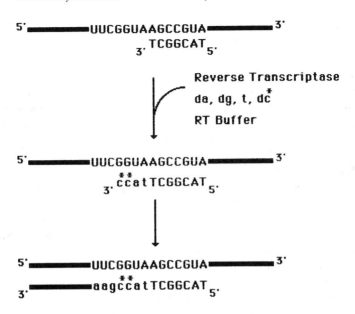

FIGURE 3.   A schematic of the manner in which nucleotides are added to the primer. *dC means that dCTP is radioactively labeled.

50 nM ^{32}P-dCTP (1 µl of 800 Ci/mmol; 1 mCi/ml)
0.12 U RNasin/µl
1.0 U reverse transcriptase/µl

Let reaction incubate at 37°C for 60 to 90 min

Of the various reaction components, the most dramatic increase in signal came with an increase in the concentration of KCl in the reaction mix to 120 mM, where the signal was increased 3- to 4-fold over that obtained with 40 mM KCl (suggested concentration for solution phase reverse transcription) (Figure 4).[21] The K$^+$ requirement of reverse transcriptase remains obscure although it has been documented since the experiments of Temin and Baltimore[22] and Goodman and Spiegelman.[23]

Dithiothreitol is present in the reaction buffer solely because RNasin, a RNase inhibitor, requires a reducing environment to be active. In control experiments, it has been found that both of these reagents can be removed from the reaction buffer with little consequence to the signal if normal protocols to limit the amount of RNase in the *in situ* transcription buffer are used in preparation of the reaction buffer. An important caveat to this, however, occurs when using ^{35}S-labeled triphosphate in the cDNA synthesis reaction, where a DTT concentration of 10 mM is recommended to limit the formation of disulfide bonds between the labeled nucleotide and between the proteins in the tissue section.

K$^+$ (mM)

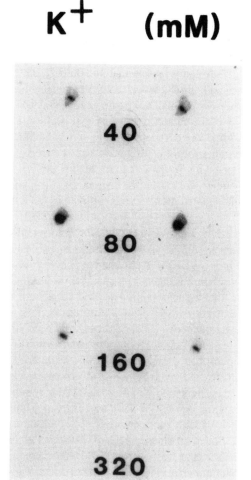

FIGURE 4.   The IST reaction buffer is identical in all cases except for indicated variations in the concentration of KCl for each reaction. The plane of sectioning through these pituitaries is such that the intermediate lobe appears as either a single or a double line through the middle of the section. At a high salt concentration (320 mM), the activity of the reverse transcriptase is inhibited. Not shown here is the optimal concentration of 120 mM which is presented elsewhere.[21]

While increasing the concentration of reverse transcriptase in the reaction does result in an increase in signal, the suggested 1 U/µl is used because larger quantities result in substantially increased costs. A reaction time of 60 to 90 min results in a good signal with minimal effect on background.

An initial worry concerning this procedure was the small amount of labeled dCTP that is included in the reaction buffer. It is conceivable that such low levels of this deoxynucleotide triphosphate may be limiting the reaction. Hypothetically, the *in situ* transcription generated cDNA might not be very long, thus limiting the signal. In experiments to address this question, labeled dCTP was diluted with varying amounts of cold dCTP. The signal is greatly reduced when 10 times more cold than labeled dCTP were used. Full length cDNA synthesis would result in a cDNA that is approximately 500 bases in length. Since the *in situ* transcription reaction products can easily attain a length of 500 bases under the conditions just described, in this experiment, one may be simply diluting out the specific activity of the signal. If a primer that is closer to the 3'-end of the mRNA is used to prime longer transcripts then it may be necessary to dilute the labeled dCTP to produce the longer transcripts and hence a stronger signal. The question of whether higher specific activity cDNA transcripts or longer transcripts will yield better *in situ* transcription signals is still unanswered. Additionally, any nucleotide, not just dCTP, can be utilized as the labeled substrate.

C. Washing Conditions Following the *In Situ* Transcription Reaction

Thorough washing of the sections after the *in situ* transcription reaction is important because it is necessary to eliminate unincorporated deoxynucleotide triphosphates. These triphosphates are charged molecules that will "stick" nonspecifically to the tissue section resulting in a high background. The conditions adopted to reduce this type of background include washing in $2\times$ SSC for 30 min at room temperature and $0.5\times$ SSC, 2.5 mM NaPP$_i$ for 8 h at 42°C. The NaPP$_i$ is included because it competes for nonspecific phosphate interaction of unincorporated triphosphates. Additionally, if the nucleotide is ^{35}S-labeled then 10 mM DTT should be used in the washing buffers to limit disulfide formation. The required time of washing may vary from tissue to tissue and depend upon the amount of protein within the tissue. This washing time should be empirically determined for each experimental situation. After washing is completed the sections should be dipped 10 times in each of a series of graded alcohols: H_2O/ 300 mM ammonium acetate, 50% ethanol/300 mM ammonium acetate, 70% ethanol/300 mM ammonium acetate, 90% ethanol/300 mM ammonium acetate, and 98% ethanol/300 mM ammonium acetate. After the sections are air dried they are ready for either film autoradiography or emulsion autoradiography.

III. THE USE OF *IN SITU* TRANSCRIPTION TO LOCALIZE mRNA MOLECULES IN TISSUE SECTIONS

The *in situ* transcription procedure was developed by examining POMC gene expression in the rat pituitary. This system was chosen because it has been well characterized by a number of investigators and hence mRNA localization as well as changes in mRNA level can be predicted. *In situ* hybridization for POMC in the rat pituitary has demonstrated that POMC mRNA is present in the melanotrophs of the intermediate lobe as well as in the corticotrophs of the anterior lobe. Dependent upon the type of probe used in *in situ* hybridization studies, film

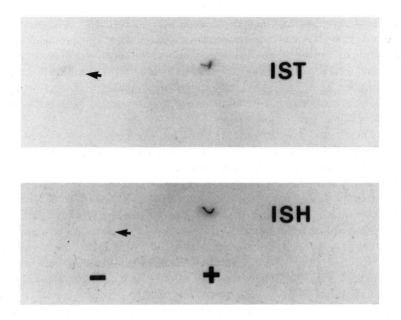

FIGURE 5. The differences in exposure time necessary to obtain a POMC hybridization signal using *in situ* hybridization and *in situ* transcription are shown here. The (+)ISH section were hybridized with a 36-base POMC oligonucleotide labeled to a specific activity of 1×10^6 cpm/µg of DNA and exposed to film for 65 h at room temperature. The (+)IST section was processed through the *in situ* transcription procedure described in the chapter and exposed to film for 15 min at room temperature. The (–) sections (indicated by the arrow) were taken through the *in situ* hybridization and *in situ* transcription procedures as described,[25] but differ from the (+) sections in that they did not have any oligonucleotide added in the hybridization step.

autoradiography of the specific ^{32}P-labeled signal takes between 2 h and 1 week to be nicely visible.

A comparison of *in situ* hybridization with *in situ* transcription shows that comparable signals are generated by varying the exposure time of the film autoradiograms such that the *in situ* hybridized section was exposed to film for 96 h while the *in situ* transcription section was exposed to film for 10 min (Figure 5). While in this comparison the POMC oligonucleotide was not radioactively labeled to a high specific activity (approximately 1×10^6 cpm/µg of DNA), the most important feature of this figure is the comparison of the background signals obtained when oligonucleotide is eliminated from the hybridization mix. The background is readily apparent in the section that did not receive any exogenously added primer. This background is not due to nonspecific sticking of labeled nucleotides, as evidenced by the fact that comparable sections taken through the *in situ* transcription procedure with no reverse transcriptase added result in much lower background. Additionally, as has been shown in Tecott et al.,[9] cDNA is synthesized in tissue sections which have gone through a standard *in situ* transcription reaction in the absence of primer in the hybridization mix. The background comes from reverse transcriptase utilizing endogenous double-stranded primer template complexes as sites for the initiation of cDNA synthesis. This endogenous background suggests that the intensity of the POMC *in situ* transcription signal may come from a competition for priming sites between the endogenous primer-template complexes and those formed by the hybridization of added primer with its mRNA.

In many cases, dependent upon the abundance of the mRNA of interest, this background will be of little consequence. In other instances, however, this background can obscure the specific signal.

In an effort to increase the signal-to-noise ratio of the *in situ* transcription procedure we have attempted to decrease the amount of endogenous priming using several methods. One of these methods, which has proven to be useful, is the denaturation of the endogenous primer-template complexes. Denaturation of these complexes must be accomplished prior to the hybridization of the specific primer. This denaturation is accomplished by dipping the tissue section in 0.5× SSC at 85°C for 5 min, followed by cooling to room temperature and addition of the hybridization mix. While it is most likely that the increase in specific signal is due to a decrease in the amount of endogenous priming, we cannot rule out the possibility that accessibility of POMC mRNA to primer and reverse transcriptase is increased by this heating step which would result in an increase in the ratio of POMC primer to endogenous primer.

When utilizing the parameters discussed to this point in the chapter, the exposure time for film autoradiography of the POMC *in situ* transcription signal using ^{35}S-labeled dCTP as labeled substrate can be reduced to 30 s (Figure 6). Such rapid exposures allow quick decisions to be made concerning the presence of mRNA within a tissue as well as whether the mRNA of interest is hormonally responsive in the tissue being examined. The dopamine responsiveness of POMC mRNA is readily apparent from a 10-min exposure of pituitary sections cut from animals that had been treated with haloperidol (dopamine antagonist)[24,25] and bromocryptine (dopamine agonist)[24] (Figure 7).

Another method for reducing background that provides approximately a threefold reduction in background is to perform a pre-*in situ* transcription reaction in which the tissue sections are taken through an *in situ* transcription reaction prior to hybridization of the specific primer. This prereaction is done in the presence of dideoxynucleotide triphosphates which when incorporated into a growing cDNA strand will terminate cDNA synthesis (Figure 8). The idea behind this background reduction strategy is to terminate or block all endogenous priming sites so that the specific primer will not have to compete with these sites for reverse transcriptase. Although this is not a large reduction in background, it is anticipated that variations on this theme will reduce background significantly.

Gene expression at the cellular level can also be explored with this technique. As shown in Figure 9, a ^{35}S-POMC *in situ* transcription pituitary section has been

FIGURE 6.   POMC IST signal obtained after a 30-s exposure using ³⁵S-labeled dCTP as the label. The required time of exposure was reduced significantly (from 10 min to 30 s) by pre-denaturing the section, followed by using optimized IST conditions described in the chapter.

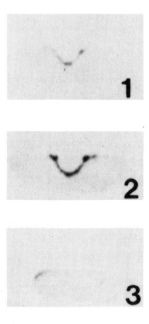

FIGURE 7.   These film autoradiograms show that the IST technique can be used to detect regulatory changes in the amount of POMC mRNA in the rat pituitary. Section 1 was taken from a rat that was given control injections of saline once a day for 4 d. Section 2 was from an animal treated with haloperidol for 4 d (3 mg/kg/d). Section 3 was from an animal treated with bromocryptine using the same injection regimen. All sections were treated equivalently in the *in situ* transcription protocol with a film exposure time of 10 min. The difference in intensity of intermediate lobe signal reflects differences in the amount of POMC mRNA as a function of these drug manipulations.

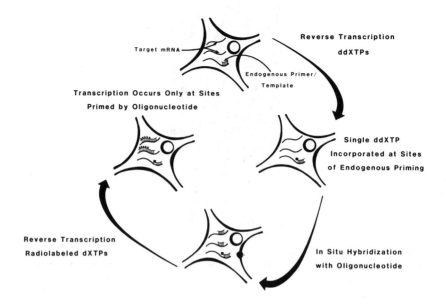

FIGURE 8.   Illustrated here is the strategy behind the addition of dideoxynucleotide triphosphates to the pre-*in situ* transcription reaction.

dipped in NTB-2 photographic emulsion and developed in D-19 developer. The exposure time to obtain this signal was 5 h (Figure 9). It is important to use test sections exposed for differing times so that the appropriate emulsion exposure times for these strong signals can be determined to prevent overexposure. In this figure the intermediate lobe is clearly visible with reasonable cellular localization of the grains. The background resulting from endogeneous priming sites is apparent from the large number of grains over all of the cells visible in this field of anterior lobe.

While the *in situ* transcription technique was developed by examining the pituitary POMC system, it can also be used effectively to examine mRNA localization in the brain. In Figure 10, a film autoradiogram of a ^{35}S-vasopressin *in situ* transcription signal in the mouse brain is shown. The suprachiasmatic nucleus (SCN) shows a signal corresponding to the presence of vasopressin mRNA within cells of this nucleus. Vasopressin mRNA localization has been documented in this region previously,[26-28] yet the exposure time was significantly longer than that obtainable by use of the *in situ* transcription technique, which was 2 h.

With increasing magnification it is clear that there is a strong signal over the SCN, yet, because of the overexposure, the background, which can be a confusing parameter of this technique, can be seen (Figure 11). In going to the cellular level the signal-to-noise ratio becomes readily apparent. The bright-field image shown in Figure 12 is overexposed yet the tissue was exposed to emulsion

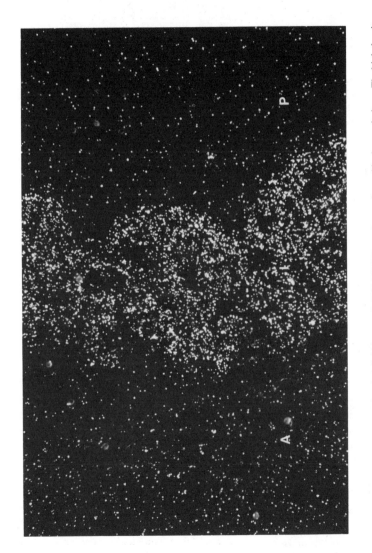

FIGURE 9. This photograph is a darkfield image of a POMC IST reaction performed in the rat pituitary. The high density signal (the band running down the middle of the field) is the intermediate lobe (I) while the anterior lobe (A) is to the left and the posterior lobe (P) is to the right of this tissue.

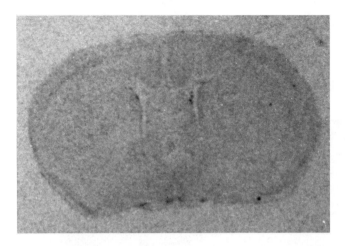

FIGURE 10. *In situ* transcription of mouse brain with arginine vasopressin C-terminal peptide cDNA oligonucleotide primer. The brain slice was prepared as described in text, and hybridized for 18 h with 500 ng of a 36-mer oligonucleotide complementary to the 3'-region of the vasopressin mRNA. IST was performed as in the text with ^{35}S-dCTP as label. Section was autoradiographed for 2 h on Kodak XAR film. Localization of signal in the suprachiasmatic and supraoptic nuclei is clearly visible.

for only 5 h. Clearly, the emulsion autoradiographic exposure time can be reduced significantly in these experiments.

IV. ELECTROPHORETIC ANALYSIS OF *IN SITU* TRANSCRIPTION-DERIVED cDNA TRANSCRIPTS

After the *in situ* transcription signal has been visualized, the cDNA transcripts can be removed from the section by treating the sections with a reagent that destabilized the hydrogen bonding of the cDNA to the mRNA. Success has been achieved using 0.2 N sodium hydroxide, 0.2 N potassium hydroxide, and 4 *M* guanidine hydrochloride. Sodium hydroxide is routinely used because it is easy to neutralize and the sodium counterion is the same as that used in the precipitation of the cDNA. The procedure involves dispersion of the tissue section with 0.2 N NaOH followed by neutralization with 1 *M* Tris-HCl pH 7.0. Ten micrograms of glycogen are added to the sample, followed by phenol-chloroform extraction and ethanol precipitation in the presence of 500 m*M* NaCl. The precipitate is washed several times in 95% ethanol and if necessary (i.e., if the salt pellet is large) ethanol precipitated a second time. The cDNA as such is ready for gel electrophoresis. The gels that are normally used for this analysis are 5% acrylamide-urea gels[29] standardly used in DNA sequencing.

The electrophoretic analysis of POMC *in situ* transcription derived transcripts

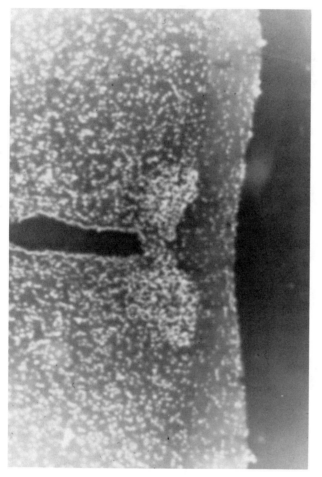

FIGURE 11. Darkfield image of silver grain localization for vasopressin message in the mouse suprachiasmatic nucleus. Brain section from Figure 11 was dipped in Kodak NTB-2 photographic emulsion, exposed for 5 h, and developed in Kodak D-19 developer. Localization of signal in suprachiasmatic nucleus at base of third ventricle is apparent. Additional grains are indicative of substantial backgrounds produced with the IST technique using nonoptimized conditions.

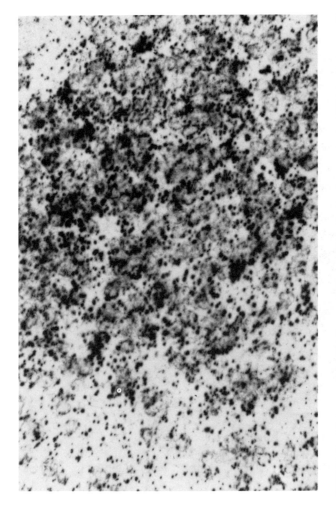

FIGURE 12. High power brightfield image of silver grain localization for vasopressin mRNA in mouse suprachiasmatic nucleus. Section from Figure 12 was hematoxolin stained. Perinuclear localization of silver grains in individual cells is apparent, demonstrating the usefulness of the IST technique in studies of gene expression in heterogeneous brain nuclei.

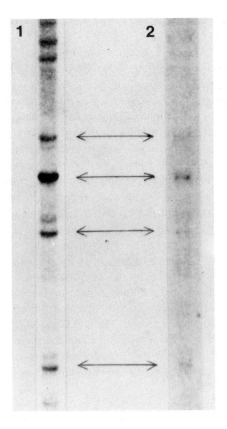

FIGURE 13. In lane 1 of this autoradiogram is the IST-generated cDNA banding pattern obtained from POMC-oligonucleotide primed pituitary ISTs using radiolabeled deoxycytosine triphosphate (limiting dCTP). In lane 2 is the banding pattern obtained from IST cDNAs which were primed with a radiolabeled POMC oligonucleotide with all of the nonlabeled deoxynucleotide triphosphates present at 100 μM levels. The arrows point out the banding pattern resulting from the two reactions.

is shown in Figure 13. There is a distinctive banding pattern that corresponds to specific termination of cDNA synthesis along the POMC mRNA molecule. The cause of termination is unclear at present, yet, as might be expected, it does not result from limiting substrate. This was determined by comparing the electrophoretic pattern of cDNA synthesized using P^{32}-labeled dCTP in the standard protocol and cDNA synthesized using the same oligonucleotide as a primer with the modification that the primer is prelabeled at the 5′-end with P^{32} (using T_4 polynucleotide kinase and γ-labeled ATP) and a high concentration of dCTP in the transcription reaction. If the banding pattern corresponds to limiting amounts of substrate, then a banding pattern would not be expected in the gel lane where a high concentration of dCTP is present. In Figure 13 the same banding pattern can be seen in both cases, thus proving that limiting dCTP is not the cause of cDNA termination.

Electrophoretic analysis of *in situ* transcription-derived cDNAs can yield interesting information concerning several aspects of cellular functioning as well as a way of confirming specificity of hybridization signals. While *in situ* transcription can be used to detect hormonally induced changes in mRNA levels (Figure 8), the intensity of certain cDNA transcripts upon gel electrophoresis also appears to be reflective of the amount of secretion that occurs for the peptide derived from a given species of mRNA. Preliminary evidence, using AtT$_{20}$ cells, suggests that the higher molecular weight POMC derived *in situ* transcription bands appear to decrease in intensity upon stimulation of hormonal secretion.[30] While the reason for these intensity changes is unclear, such changes may be reflective of alterations in the amount of translation that are occurring for that species of mRNA.[31]

There is precedent for translational control in a number of systems with recent work by Casey et al.[31] in which they describe the existence of iron-responsive regulatory elements which may be involved in controlling aspects of mRNA translation. Experiments are currently underway to determine the cause of alterations in the intensity of the banding pattern resulting from differences in hormonal manipulation of the cells. The specificity of hybridization signals generated in *in situ* hybridization and *in situ* transcription analysis can be determined using multiple primers that can be specifically modified in the *in situ* transcription technique and are discussed in detail in Zangger et al.[21] and Tecott and Eberwine.[32]

Since *in situ* transcription-derived cDNAs can be isolated, they can also be cloned. Cloning of such small quantities of cDNA can be accomplished using standard cDNA cloning techniques. It is necessary to make sure that either highly competent bacteria (10^9 colonies/µg) for plasmid vectors or lambda cloning vectors are utilized so that a maximal yield of colonies (plaques) per nanogram of starting cDNA is obtained. While this strategy can be used for relatively high abundance mRNAs (e.g., POMC mRNA in the rat pituitary) it is difficult to clone lower abundance mRNAs using these techniques. To circumvent this problem we have developed a technique that allows us to amplify the *in situ* transcription-derived cDNAs to larger, easily clonable amounts.[33] Briefly, the cDNAs from *in situ* transcription, primed with oligo-T (so that all polyadenylated mRNAs are copied into cDNA), are extended at the 3′-end with deoxyguanosine triphosphate using terminal deoxynucleotidyl transferase. This 3′-polyG tail provides a priming site for oligo-dC to act as a primer for the synthesis of the complementary strand of cDNA. The resultant double-stranded DNA has an oligo-dC priming site on one end and an oligo-T priming site on the other side, thereby providing two "specific" sites that can be used in the polymerase chain reaction[34] to amplify the internal cDNA. The ability to amplify small amounts of cDNA isolated from highly defined brain regions (from the *in situ* transcription procedure) permits the synthesis of very specific cDNA libraries which are useful in defining mRNAs of regionally specific function.

V. USE OF *IN SITU* TRANSCRIPTION FOR THE NONRADIOACTIVE LOCALIZATION OF mRNA MOLECULES

The *in situ* transcription procedure requires the use of radioactive substrates which often increase autoradiographic background and at some institutions are difficult to dispose of. We have adapted the *in situ* transcription technique for use in nonradioactive detection paradigms by incorporating biotinylated-dUTP into the transcription buffer, thus producing biotinylated cDNA. The biotinylated cDNA has been detected using an enzymatic alkaline phosphatase reaction developed by Unger et al.[35] (Figures 14A and 14B). This was done by binding the avidin of avidin-conjugated alkaline phosphatase to the biotin groups. The attached alkaline phosphatase can then be used to catalyze the conversion of nitro-blue tetrazolium chloride (NTB) and 5-bromo-4-chloro-3-indoyl phosphate *p*-toluidine salt (BCIP) to dark blue formazan precipitate. The formazan precipitate from this reaction, also known as the McGadey reaction,[36] is easily visible using light microscopy. Given the success of using the avidin-biotin detection method, it should also be possible to use FITC-labeled avidin for fluorescence detection of the biotinylated cDNAs.[6] As new nonradioactive detection strategies utilizing different types of modified nucleotides (that are not biotinylated) become available, they may be easily adaptable for use in the *in situ* transcription procedure.

The ability to detect biotinylated cDNA within a tissue section suggests that colocalization of mRNAs (using the *in situ* transcription procedure) with peptides (using immunohistochemistry techniques)[37,38] is possible using simple, rapid, and straightforward techniques. Such information would permit the rapid determination of alterations in the amount of peptide in conjunction with changes in the level of mRNA within individual cells.

VI. CONCLUSIONS

As discussed in this chapter, the use of *in situ* transcription allows the localization of mRNAs, electrophoretic analysis of *in situ* transcription-derived cDNA transcripts, regional cloning of localized cDNAs, and, potentially, examination of the translational state of a specific mRNA. Additionally, *in situ* transcription combined with polymerase chain reaction should permit the cloning of cDNAs from small amounts of precious tissue. While there is much work to be done to reduce endogenous priming and to optimize various aspects of the *in situ* transcription procedure, the potential use of *in situ* transcription technology is to examine chronic and acute changes in mRNA levels in different

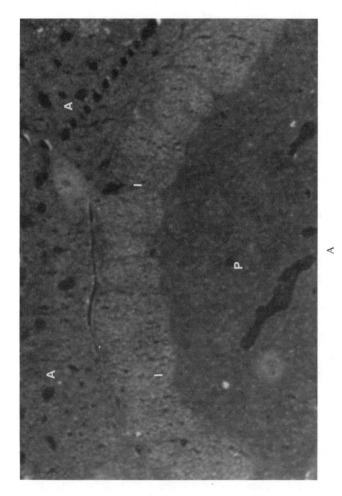

FIGURE 14. The use of biotin-labeled dUTP as a substrate in the IST reaction. Photo 14A shows the positive signal obtained over the intermediate lobe (indicated with I). Photo 14B shows the control in which the IST reaction was performed using the same reaction conditions with the addition of 200 µ*M* TTP to the reaction mix. The TTP will compete with the biotinylated dUTP, thus significantly reducing the specific signal. The A indicates the position of the anterior lobe while P indicates the position of the posterior lobe.

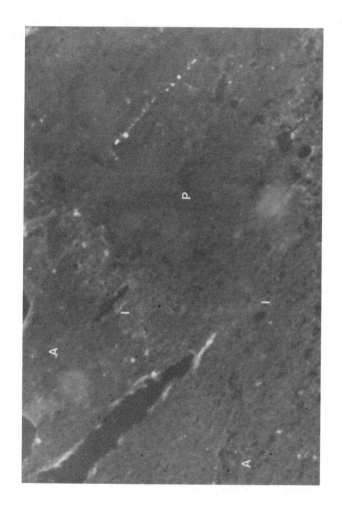

FIGURE 14B.

pharmacological and behavioral states such as opioid tolerance, circadian rhythmicity, and changes related to various disease states.

ACKNOWLEDGMENTS

This work was supported by grants MH23861, DA0501, N00014-86-K-0251, and MH09099. The continuing encouragement of Jack Barchas is greatly appreciated. David Newell and Andrew Hoffman have collaborated with us on the issue of whether the *in situ* transcription technique can be used to determine the translational state of mRNA. The suggestion of predenaturing the sections prior to primer hybridization was made by Susan Fox. We would like to thank Lucy Movahed and Lawrence Weiss for bringing the avidin-alkaline phosphatase development procedure to our attention.

REFERENCES

1. **Coghlan, J. P., Penschow, J. D., Fraser, J. R., Aldred, P., Haralambidis, J., and Tregear, G. W.,** Location of gene expression in mammalian cells, in *In Situ Hybridization: Applications to Neurobiology,* Valentino, K., Eberwine, J., and Barchas, J., Eds., Oxford University Press, New York, 1987, chap. 2.
2. **Pintar, J. E. and Lugo, D. I.,** Localization of peptide hormone gene expression in adult and embryonic tissues, in *In Situ Hybridization: Applications to Neurobiology,* Valentino, K., Eberwine, J., and Barchas, J., Eds., Oxford University Press, New York, 1987, chap. 10.
3. **Brahic, M. and Haase, A. T.,** Detection of viral sequences of low reiteration frequency by *in situ* hybridization, *Proc. Natl. Acad. Sci. U.S.A.,* 75, 6125, 1978.
4. **Angerer, L. M. and Angerer, R. C.,** Detection of poly A+ RNA in sea urchin eggs and embryos by quantitative *in situ* hybridization, *Nucl. Acids Res.,* 9, 2819, 1981.
5. **Gee, C. E., Chen, C. L., Roberts, J. L., Thompson, R., and Watson, S. J.,** Identification of proopiomelanocortin neurons in rat hypothalamus by *in situ* cDNA-mRNA hybridization, *Nature,* 306, 374, 1983.
6. **Singer, R. H., Lawrence, J. B., and Raschtchian, R. N.,** Toward a rapid and sensitive *in situ* hybridization methodology using isotopic and nonisotopic probes, in *In Situ Hybridization: Applications to Neurobiology,* Valentino, K., Eberwine, J., and Barchas, J., Eds., Oxford University Press, New York, 1987, chap. 4.
7. **Shivers, B. D., Schachter, B. S., and Pfaff, D. W.,** *In situ* hybridization for the study of gene expression in the brain, *Methods Enzymol.,* 124, 497, 1986.
8. **Baltimore, D. and Smoler, D. F.,** Primer requirement and template specificity of the DNA polymerase of RNA tumor viruses, *Proc. Natl. Acad. Sci. U.S.A.,* 68, 1507, 1971.
9. **Houts, G. E., Miyagi, M., Ellis, D., Beard, D., and Beard, J. W.,** Reverse transcriptase from avian myeloblastosis virus, *J. Virol.,* 29, 517, 1979.
10. **Tecott, L. H., Barchas, J. D., and Eberwine, J. H.,** *In situ* transcription: specific synthesis of complementary DNA in fixed tissue sections, *Science,* 240, 1661, 1988.
11. **Drouin, J., Chamberland, M., Charron, J., Jeanotte, L., and Nemer, M.,** Structure of the rat pro-opiomelanocortin (POMC) gene, *FEBS Lett.,* 193, 54, 1985.

12. **Eberwine, J. H.,** Glucocorticoid and Corticotropin Releasing Hormone Regulation of Pro-Opiomelanocortin Gene Expression, doctoral dissertation, Columbia University, New York, 1984.

13. **Schmale, H., Heinsohn, S., and Richter, D.,** Structural organization of the rat gene for the arginine vasopressin-neurophysin precursor, *EMBO J.,* 2, 763, 1983.

14. **Lewis, M. E., Sherman, T. G., and Watson, S. J.,** *In situ* hybridization histochemistry with synthetic oligoneucleotides: strategies and methods, *Peptides,* 6(Suppl. 2), 75, 1985.

15. **Angerer, L. M., Stoler, M. H., and Angerer, R. C.,** *In situ* hybridization, with RNA probes: an annotated recipe, in *In Situ Hybridization: Applications to Neurobiology,* Valentino, K., Eberwine, J., and Barchas, J., Eds., Oxford University Press, New York, 1987, 42.

16. **Lewis, M. E., Sherman, T. G., Burke, S., Akil, H., Davis, L. G., Arentzen, R., and Watson, S. J.,** Detection of proopiomelanocortin mRNA by *in situ* hybridization with an oligonucleotide probe, *Proc. Natl. Acad. Sci. U.S.A.,* 83, 5419, 1986.

17. **Wetmur, J. G. and Davidson, N.,** Kinetics of renaturation of DNA, *J. Mol. Biol.,* 31, 349, 1968.

18. **Britten, R. J., Graham, E. D., and Neufel, B. R.,** Analysis of repeating DNA sequences by reassociation, in *Methods in Enzymology,* Vol. 29, Academic Press, New York, 1974, 363.

19. **Casey, J. and Davidson, N.,** Rates of formation and thermal stabilities of RNA:DNA and DNA:DNA duplexes at high concentrations of formamide, *Nucl. Acids Res.,* 4(5), 1539, 1977.

20. **Meinkoth, J. and Wahl, G.,** Hybridization of nucleic acids immobilized on solid supports, *Anal. Biochem.,* 138, 267, 1984.

21. **Zangger, I., Tecott, L., Barchas, J., and Eberwine, J.,** *In situ* transcription: a methodological study using proopiomelanocortin gene expression in the rat pituitary as a model, submitted.

22. **Temin, H. M. and Baltimoire, D.,** RNA-directed DNA synthesis and RNA tumor viruses, *Adv. Virus Res.,* 17, 129, 1971.

23. **Goodman, N. D. and Spiegelman, D.,** Distinguishing reverse transcriptase of an RNA tumor virus from other known DNA polymerases, *Proc. Natl. Acad. Sci. U.S.A.,* 68, 2203, 1971.

24. **Hollt, V., Haarman, I., Seizinger, B. R., and Herz, A.,** Chronic haloperidol treatment increases the level of *in vitro* translatable messenger ribonucleic acid coding for the beta-endorphin/adrenocorticotropin precursor proopiomelanocortin in the pars intermedia of the rat pituitary, *Endocrinology,* 110, 1885, 1982.

25. **Chronwall, B. M., Millington, W. R., Griffin, W. S. T., Unnerstall, J. R., and O'Donohue, T. L.,** Histological evaluation of the dopaminergic regulation of proopiomelanocortin gene expression in the intermediate lobe of the rat pituitary, involving *in situ* hybridization and [³H]-thymidine uptake measurement, *Endocrinology,* 120, 1201, 1987.

26. **Sherman, T. G., Watson, S. J., Herbert, E., and Akil, H.,** The co-expression of dynorphin and vasopressin: an *in situ* hybridization and dot blot analysis of mRNAs during stimulation, in, Society for Neuroscience Annual Meeting Abstracts No. 10, 1984, 359.

27. **Uhl, G. R., Zingg, H. H., and Habener, J. F.,** Vasopressin mRNA *in situ* hybridization: localization and regulation studied with oligonucleotide cDNA probes in normal and Brattleboro rat hypothalamus, *Proc. Natl. Acad. Sci. U.S.A.,* 82, 5555, 1985.

28. **Uhl, G. R. and Reppert, S. M.,** Suprachiasmatic nucleus vasopressin messenger RNA: circadian variation in normal and brattleboro rats, *Science,* 231, 390, 1986.

29. **Maniatis, T. E., Fritsch, E. F., and Sambrook, J.,** *Molecular Cloning,* Cold Spring Harbor Laboratory, Cold Spring Harbor, NY, 1982.

30. **Tecott, L., Newell, D., Eberwine, J., and Hoffman, A.,** Blockade of ribosome binding to mRNA alters *in situ* transcription cDNAs in cultured pituitary cells, in 70th Annual Endocrine Society Meeting Abstracts, 1988.

31. **Casey, J. L., Hentze, M. W., Koeller, D. M., Caughman, S. W., Rouault, T. A., Klausner, R. D., and Harford, J. B.,** Iron-responsive elements: regulatory RNA sequences that control mRNA levels and translation, *Science,* 240(4854), 924, 1988.

32. **Tecott, L. and Eberwine, J.,** Template sequence-dependent termination of *in situ* transcription, in Society for Neuroscience Annual Meeting Abstracts No. 27, 1988b.

33. **Eberwine, J. H., Zangger, I., and Tecott, L. H.,** Consideration in the use of *in situ* transcription methodology, in *Short Course 1 Syllabus;* In Situ *Hybridization and Related Techniques to Study Cell-Specific Gene Expression in the Nervous System,* Society for Neuroscience, Toronto, 1988, 69.

34. **Saiki, R. K., Gelfand, D. H., Stoffel, S., Scharf, S. J., Higuchi, R., Horn, G. T. R., Mullis, K. B., and Erlich, H. A.,** Primer-directed enzmatic amplification of DNA with a thermostable DNA polymerase, *Science,* 239, 1988.

35. **Unger, E. R., Budgeon, H. T., Myerson, D., and Brigati, D. J.,** Viral diagnosis by *in situ* hybridization, description of a rapid simplified colorometric method, *Am. J. Surg. Pathol.,* 10(1), 1, 1986.

36. **McGadey, J.,** A tetrazolium method for non-specific alkaline phosphatase, *Histochimie,* 23, 180, 1970.

37. **Brahic, M., Haase, A. T., and Cash, E.,** Simultaneous *in situ* detection of viral RNA and antigens, *Proc. Natl. Acad. Sci. U.S.A.,* 81, 5445, 1984.

38. **Schachter, B. S.,** Studies of neuropeptide gene expression in brain and pituitary, in *In Situ Hybridization: Applications to Neurobiology,* Valentino, K., Eberwine, J., and Barchas, J., Eds., Oxford University Press, New York, 1987, chap. 6.

Chapter 7

IN SITU HYBRIDIZATION HISTOCHEMISTRY AT THE ELECTRON MICROSCOPIC LEVEL

Jean-Jacques Soghomonian

TABLE OF CONTENTS

I. INTRODUCTION

In situ hybridization histochemistry, which allows for the detection of DNA or RNA molecules in tissue sections, was initially developed by Gall and Pardue[1] and John, Burnstiel, and Jones.[2] Since then, this technique has been widely used in studies of gene mapping and gene expression in various types of mammalian and nonmammalian cells. Most of these studies have been done at the light microscopic level after hybridization of nucleic acids with radioactive or nonradioactive probes.[3]

Some have tried to extend this technique to the electron microscopic level in order to obtain better resolution in the localization of the probe and therefore gain accurate information about the subcellular localization of the nucleic acid and/ or the morphology of cells expressing the nucleic acid. The technical feasibility of visualizing DNA by *in situ* hybridization on ultrathin sections for electron microscopy was first demonstrated with oocytes of Xenopus tadpoles by Jacob et al.[4] who used tritiated RNA as a probe. In subsequent studies, different protocols have been developed according to the type of nucleic acid to be detected or to the biologic material under investigation. For instance, hybridizations have been done on chromosomal or ribosomal DNA,[5-12] viral DNA[13-15] or cellular RNA[17-21] radioactive[4,7,8,11,14,15,18-20] and/or nonradioactive probes[5-7,9,10,12,16,17,21] were used in these studies.

The protocols for electron microscopy are similar to those used for light microscopic visualization of hybrids except that they require particular treatment of the tissue and labeling procedure. Some of these requirements are contradictory with an optimal hybridization procedure and a compromise must be found between an acceptable morphology and an efficient hybridization. Here we discuss some of these technical aspects with particular emphasis on cellular *in situ* hybridization.

II. TISSUE PREPARATION

A. Fixation

Good fixation of the tissue is particularly important in electron microscopic studies of whole cells since the biological information is related to their ultrastructural integrity. The fixation is also important since it allows for the retention of the nucleic acids in their original subcellular compartment before, during, and after the hybridization procedure. Cross-linking primary fixatives such as formaldehyde and glutaraldehyde, which are the most commonly used fixatives for electron microscopic preservation of the tissue, are also efficient for retaining nucleic acids, particularly RNA molecules, in the tissue (see References 22 and 23). Glutaraldehyde gives both better RNA retention and preservation of the tissue. Glutaraldehyde has been used alone and in combination with formaldehyde in electron microscopic hybridization studies for visualizing nucleic acids in cell preparations or whole tissue slices.[4,11,18-21]

The use of these cross-linking fixatives, particularly glutaraldehyde, has the disadvantage of limiting the accessibility of the nucleic acid to the probe (see Reference 23). This may compromise the success of the hybridization, particularly when very few copies of the nucleotidic sequence to be detected are present in the tissue. Different concentrations of fixative can be tried in order to optimize the morphological preservation of the tissue while keeping a good sensitivity of mRNA detection (for instance, see References 18, 20, 21, 24). Another way to increase the efficiency of the hybridization after strong aldehyde fixation without affecting the ultrastructure is to use probes of shorter length. For instance, Angerer and Angerer[25] have reported that after glutaraldehyde fixation of sea urchin eggs, small fragments of 50 bp of tritiated poly U or poly C give a 2- to 3-fold increase in the hybridization signal on poly A+ mRNA as compared to 500 bp probes. Treatments of the tissue with proteases or detergents before the hybridization can also be done in order to increase the penetration of the probe.

Increased efficiency of hybridization with proteinase has been demonstrated on glutaraldehyde fixed urchin eggs.[25] However, on the formaldehyde-fixed myoblasts, Lawrence and Singer[26] have reported no advantage to proteolytic digestion. On the contrary, incubation with proteinase K can cause loss of more than half of the total cellular RNA. As recently reported for rat pituitary sections,[20] even for glutaraldehyde fixed tissue, pretreatment with proteinase K is detrimental to the cell's ultrastructural morphology and it should therefore be avoided for *in situ* hybridization studies at the electron microscopic level. Pretreatment of the tissue with the detergent Triton X-100 can also be used in order to facilitate the penetration of the probe. High concentrations of Triton (0.1%) have been used to enhance the signal of a 717 bp [35S]-cDNA probe for prolactin in rat pituitary fixed by a mixture of paraformaldehyde and glutaraldehyde.[20] However, such treatment also has a deleterious effect on the cellular ultrastructure. Concentrations of Triton as low as 0.02% have been reported to produce focal discontinuities in the plasma and nuclear membranes with a good preservation of the cytoplasmic organelles in human fibroblasts fixed with formalin.[16] It has been reported that a higher concentration of Triton had a more deleterious effect on the membranes.[16] Therefore, the use of protease or Triton should be avoided for electron microscopic studies if possible, although a judicious combination of these two treatments resulted in good penetration of a DNA probe with acceptable ultrastructural preservation of human fibroblasts.[16]

Primary fixation with osmium tetroxide was reported to compromise the mRNA-cDNA hybridization on thin sections of rat pituitary.[18] When used alone, this fixative has also been shown to result in poor retention of the RNA in the tissue (see Reference 23). Nevertheless, osmium tetroxide is routinely used as a secondary fixative for electron microscopy and does not seem to interfere with the labeling when used after hybridization. For instance, in the rat pituitary, hybridization of a [35S]-cDNA probe complementary to prolactin mRNA has recently been visualized after osmium postfixation.[20]

B. Pre-Embedding Techniques

In most *in situ* hybridization studies at the light microscopic level, the hybridization step is done on fixed biological material without subsequent embedding of cells or tissue. For the purpose of electron microscopy, the hybridization can also be performed prior to embedding. For tissue preparations, sectioning can be done on a vibratome and the hybridization performed by incubating the tissue slices in the hybridization mixture.[20,21] Using such thick vibratome slices (100 to 200 μm), Guitteny et al.[24] have reported complete penetration of a small oligonucleotide DNA probe in the rat hypothalamus after a very light glutaraldehyde (0.1%) fixation without detergent or protease treatment. This indicates that the thickness of the vibratome slice is not a limiting factor to the penetration of the probe. A larger DNA probe (100 to 300 bp) has also been shown to penetrate vibratome freeze-thawed slices from the rat trigeminal ganglia.[21] The extent of the penetration of the probe in tissue slices may vary in different conditions, particularly after a stronger fixation of the tissue. Some have reported successful electron microscopic visualization of cellular mRNA using a cryo-ultramicrotomy technique on thin sections of formaldehyde-fixed rat pituitary.[18] In this study, the sections were deposited on collodion-coated grids and hybridized with a radiolabeled cDNA probe for growth hormone. This technique is sensitive, but freezing the tissue can be somewhat deleterious for the ultrastructure of the cells.

After hybridization, the cells or tissue sections have to be embedded. It is important to know whether or not this can modify the stability of the hybrids and compromise their subsequent detection. As already discussed, experimental evidence indicates that postfixation with osmium tetroxide apparently does not change the stability of the hybrids.[14,16,18,20,21,24] Hybridization studies done in rat brain or monkey and human cell preparations indicate that embedding in nonhydrophilic resins, like Epon or araldite, is compatible with the visualization of the hybrids. Epon embedding medium has, for instance, been used in *in situ* detection of viral DNA in monkey cell cultures[14] or of mRNA in slices from the rat trigeminal nucleus.[21] In the rat pituitary, Guitteny et al.[24] have reported a remarkable stability of an mRNA-DNA complex during both dehydration in methanol and propylene oxide and the subsequent flat-embedding in araldite at 40°C. They nevertheless reported that with their vibratome and post-embedding protocol, the sensitivity of mRNA detection seems to be below that obtained in cryostat-cut sections. This could be due to some loss of hybrids during the embedding procedure.

C. Post-Embedding Techniques

A number of *in situ* hybridization studies have been carried out on ultrathin sections for electron microscopy after embedding in water-soluble resin. In this case, after embedding, the tissue sections are collected on formar- or collodion-coated grids and all the hybridization steps are directly performed on the sections. This technique offers some advantages. Among them is the ability to process serial thin sections with different probes or to process sections alternately for *in situ* hybridization and immunocytochemical labeling. The most important requirement is that the embedding procedure does not destroy the endogenous nucleic acids and allows their retention in their original cellular compartment.

Hydrophilic embedding media such as acrylate and methacrylate based resins have been used successfully in a number of studies. Glycol methacrylate embedded cells, for instance, were hybridized with radiolabeled DNA or RNA probes to visualize cellular RNA or DNA.[11,15,19] The low temperature embedding media Lowicryl K4M,[27] which is a polar acrylate-methacrylate resin, has been used to demonstrate hybridization of biotinylated small DNA probes (35 to 300 nucleotides) with RNA on Drosophila ovaries[17] or rat trigeminal ganglia.[21] The accessibility of nucleic acids in embedded tissue appears relatively good. Even on thin sections, however, the length of the probe may limit penetration in the tissue. On glutaraldehyde fixed and glycolmethacrylate embedded oocytes, Steinert et al.[11] have reported a higher density of silver grains, after hybridization of a ^{125}I-labeled rRNA probe, at 80°C than 70°C. This could be explained by a facilitation of the penetration of the probe in the resin at higher temperatures or by partial degradation of the probe.

III. HYBRIDIZATION

A. General Considerations

The hybridization conditions for *in situ* hybridization histochemistry are mainly determined by the probe used and the nucleic acid to be detected. The adaptation of this technique for electron microscopic visualization has to take into account an eventual deterioration of the ultrastructural morphology during the hybridization step. The high temperature and/or long incubation time required for the hybridization may result in poor cellular morphology. The possibility of decreasing the temperature by hybridizing in higher concentrations of formamide is limited since this treatment also has a deleterious effect on the ultrastructural morphology.[17] Some studies have, however, reported acceptable ultrastructural preservation after incubation with formamide.[17,20,21] Acetylation of the tissue, which is commonly used in *in situ* hybridization with light microscopy to decrease nonspecific background,[28] has a detrimental effect on the tissue at the electron microscopic level.[17] This step can be avoided by preincubation of the tissue in the hybridization mix (see Reference 29). The type of label used for the probe also affects the sensitivity and resolution of hybrid visualization at the electron microscopic level. Probes used for *in situ* hybridization can be labeled either by radioactive or nonradioactive markers.

B. Radioactive Probes

As for light microscopic studies, the use of radiolabeled probes offers many advantages: sensitivity of detection, no interference with the hybridization reaction, and a suitability for quantification. The labeling of DNA or RNA probes is routinely achieved by the incorporation of radioactive nucleotides. This can be done by nick translation, random priming, or *in vitro* transcription in the case of RNA molecules (see Reference 3). For oligonucleotide probes, high specific activities are obtained after terminal incorporation of radioactive nucleotides.[30] Three radioisotopes are commonly used for probe labeling, ^{32}P, ^{35}S, or ^{3}H, with respective β-ray energies of 1.71, 0.167, and 0.0186 MeV. These radioisotopes have been used for the electron microscopic detection of nucleic acid hybrids in various tissue and cell preparations. Steinert et al.[11] have also used ^{125}I-labeled rRNA for the electron microscopic visualization of the amplified ribosomal RNA during *Xenopus laevis* oocyte maturation. Because of its low emission energy, tritium allows for a good determination of the emitting source but needs longer time of exposure than other radioisotopes, and, because of its higher energy, ^{35}S is usually preferred. In a comparative study using these isotopes, Li et al.[8] have reported that all three are approximately equally effective with respect to the amount of signal detected and the ratio of specific to nonspecific rDNA labeling on human metaphase chromosomes. However, they found that ^{3}H-labeled probes were superior in labeling rate (i.e., the specific labeling per day of radioautographic exposure and per metaphase). Using a cDNA probe labeled by nick translation with ^{3}H-dCTP and ^{3}H-TTP or ^{35}S-dCTP, Morel et al.[18] found that ^{35}S was a good compromise between a shorter exposure time and an acceptable resolution to visualize growth hormone mRNA in the nucleus and cytoplasm of somatotropic cells in rat pituitary. In the nucleus, the silver grains were found associated with the euchromatin and nuclear membrane but the low definition obtained with their technique using frozen thin sections did not allow a precise localization of labeling in the subcellular compartments. Using a ^{35}S-prolactin cDNA, Tong et al.[20] have recently been able to demonstrate preferential association of the labeling with the rough endoplasmic reticulum and the nucleus of two prolactin-producing cell types in the rat pituitary.

C. Nonradioactive Probes

The advantages of using nonradioactive probes are the same for light or electron microscopy. They do not necessitate long radioautographic exposure time or specific precautions during their handling, and the resolution obtained with these probes can be higher than that obtained with radioactive probes. The labeling of the probe can be done either by an enzymatic incorporation of labeled nucleotides into the nucleic acid or by a chemical attachment of the marker to the nucleic acid.

Biotylinated probes have been used in a number of electron microscopic studies.[5-7,9,10,16,17,21] These probes can be obtained by nick translation of a DNA and incorporation of a biotin-substituted deoxy-UTP (bio-dUTP) or by *in vitro* synthesis of a cRNA in presence of bio-UTP.[31] It has been shown that the cRNA synthesis reaction in the presence of bio-UTP is about 10% as efficient as a parallel reaction with unlabeled UTP.[7] Manning et al.[10] have also used a covalent attachment of biotin on RNA by a cytochrome bridge to visualize DNA sequences on *Drosophila melanogaster* chromosomes. Biotinylated probes can be detected indirectly with an antibiotin antibody and a second antibody conjugated to peroxidase or to gold particles or by a protein A complex. Avidin molecules coupled with peroxidase or gold particles were also used to detect biotin-labeled DNA after hybridization with viral DNA[16] and cellular mRNA.[21] Nonspecific sticking of avidin to cells may limit the use of this method, however (see Reference 23). Using acrylate-methacrylate embedded Drosophila ovaries, Binder et al.[17] visualized DNA hybrids with a biotin-antibiotin-protein A-gold system and compared the labeling with tritium-labeled probes. Both probes gave the same pattern of distribution, but the radioautographic detection system was inferior in terms of spatial resolution and structural preservation. However, with the biotin-labeled probe, sections exhibited contamination with semi-electron dense specks and with an amorphous layer obscuring morphological detail. This was resolved by replacing posthybridization washes with formamide by phosphate buffer.[17]

N-Acetoxy-2-acetylamino-fluorene (N-AcO-AAF)-labeled DNA and RNA probes[32] have been used for the electron microscopic visualization of hybrids on human metaphase chromosomes.[5] These probes were detected with anti-N-AcO-AAF antibodies and antibody-peroxidase or antibody-gold complexes. Both staining procedures allowed for a positive localization of ribosomal genes but only the peroxidase reaction was suitable for the visualization of a single copy sequence. Poly(dA) tails linked to a cloned Drosophila DNA and hybridized with poly(dT) tails labeled by gold spheres were used as a probe for the electron microscopic mapping of Drosophila polytene chromosomes.[12] This gold method was reported to give a low background and a high resolution.

Colloidal gold appears to give a better resolution but lower sensitivity than peroxidase since the activity of the enzyme results in the accumulation of reaction products rendering the labeled sites rather large. The advantage of using gold particles as a detection system was illustrated by Webster et al.[21] on mRNA in rat Schwann cells hybridized with biotylinated cDNA probes and by Hutchison et al.[7] and Manuelidis et al.[9] on mouse satellite DNA hybridized with DNA- or RNA-biotinylated probes. Also using a gold detection system, Wolber et al.[16] described the localization of a viral cDNA probe at the interface of electron-dense and electron-lucent regions of nuclear inclusions in cultured human fibroblasts. The low sensitivity of colloidal gold as compared to peroxidase could be due to a limited penetration of the 20 nm gold particles. This can be circumvented by the use of smaller (5 nm) particles.[33] Alternatively, the low sensitivity could be related to the repulsion of gold particles and chromatin which are both negatively charged. This can be minimized by enlarging the distance between them through application of a multistep method or by using biotin-nucleotides with longer linker arms so that the biotinyl group is further from the hybrid. Using a biotinylated-protein A-gold method on Drosophila follicle cells, Binder et al.[17] extrapolated a value of about one gold particle per ten rRNA molecules which indicated a rather good sensitivity.

IV. CONCLUSION

The combination of *in situ* hybridization and electron microscopy has been done in a relatively small number of studies. The search for an optimal ultrastructural preservation of the tissue may often limit the sensitivity and specificity of hybridization. However, the variety of methods used successfully indicate that different protocols can be adapted depending on the type of nucleic acid to be detected, the probe used, and the biological specimen studied. New methodological developments are likely to allow further refinement of *in situ* hybridization at the electron microscopic level.

ACKNOWLEDGMENTS

The author wishes to thank Dr. M. F. Chesselet for her support and helpful discussions during the preparation of this manuscript and Dr. M. Mercugliano for critical reading. J. J. S. is a recipient of a fellowship from the Pharmaceutical Manufacturers Association Foundation (PMAF).

REFERENCES

1. **Gall, J. G. and Pardue, M.,** Formation and detection of RNA-DNA hybrid molecules in cytological preparations, *Proc. Natl. Acad. Sci. U.S.A.,* 63, 378, 1969.
2. **John, H. A., Birnstiel, M. L., and Jones, K. W.,** RNA-DNA hybrids at the cytological level, *Nature,* 223, 582, 1969.
3. **Pardue, M. L.,** *In situ* hybridization, in *Nucleic Acid Hybridization,* Hames, B. D. and Higgins, S. J., Eds., IRL Press, Oxford, 1985, chap. 8.
4. **Jacob, J., Todd, K., Birnstiel, M. L., and Bird, A.,** Molecular hybridization of ³H-labelled ribosomal RNA with DNA in ultrathin sections prepared for electron microscopy, *Biochem. Biophys. Acta,* 228, 761, 1971.
5. **Cremers, A. F. M., Jansen in de Wal, N., Wiegant, J., Dirks, R. W., Weisbeek, P., van der Ploug, M., and Landegent, J. E.,** Non-radioactive *in situ* hybridization. A comparison of several immunocytochemical detection systems using reflection-contrast and electron microscopy, *Histochemistry,* 86, 609, 1987.
6. **Ferguson, D. J. P., Burns, J., Harrison, D., Jonasson, J. A., and McGee, J. O. D.,** Chromosomal localization of genes by scanning electron microscopy using *in situ* hybridization with biotinylated probes: Y chromosome repetitive sequences, *Histochemistry,* 18, 266, 1986.
7. **Hutchison, N. J., Langer-Safer, P. R., Ward, D. C., and Hamkalo, B. A.,** *In situ* hybridization at the electron microscope level: hybrid detection by radioautography and colloidal gold, *J. Cell Biol.,* 95, 609, 1982.

8. **Li, C.-B., Wu, M., Margitich, I. S., and Davidson, N.,** Gene mapping on human metaphase chromosomes by *in situ* hybridization with ^{3}H, ^{35}S, and ^{32}P labeled probes and transmission electron microscopy, *Chromosoma,* 93, 305, 1986.

9. **Manuelidis, L., Langer-Safer, P. R., and Ward, D. C.,** High resolution mapping of satellite DNA using biotin-labeled DNA probes, *J. Cell Biol.,* 95, 619, 1982.

10. **Manning, J. E., Hershey, N. D., Broker, T. R., Pellegrini, M., Mitchell, H. K., and Davidson, N.,** A new method of *in situ* hybridization, *Chromosoma,* 53, 107, 1975.

11. **Steinert, G., Thomas, C., and Brachet, J.,** Localization by *in situ* hybridization of amplified ribosomal DNA during *Xenopus laevis* oocyte maturation (a light and electron microscopic study), *Proc. Natl. Acad. Sci. U.S.A.,* 73, 833, 1976.

12. **Wu, M. and Davidson, N.,** Transmission electron microscopic method for gene mapping on polytene chromosomes by *in situ* hybridization, *Proc. Natl. Acad. Sci. U.S.A.,* 78, 7059, 1981.

13. **Croissant, O., Dauguet, C., Jeanteur, P., and Orth, G.,** Application de la technique d'hybridation moléculaire *in situ* à la mise en évidence au microscope électronique, de la réplication végétative de l'ADN viral dans les papillomes provoqués par le virus de Shope chez le Lapin cottontail, *C. R. Acad. Sci. Paris,* 274, 614, 1972.

14. **Geuskens, M. and May, E.,** Ultrastructural localization of SV40 viral DNA in cells, during lytic infection, by *in situ* molecular hybridization, *Exp. Cell Res.,* 87, 175, 1974.

15. **Jacob, J., Gillies, K., Macleod, D., and Jones, K. W.,** Molecular hybridization of mouse satellite DNA-complementary RNA in ultrathin sections prepared for electron microscopy, *J. Cell. Sci.,* 14, 253, 1974.

16. **Wolber, R. A., Beals, T. F., Lloyd, R. V., and Maassab, H. F.,** Ultrastructural localization of viral nucleic acid by *in situ* hybridization, *Lab. Invest.,* 59, 144, 1988.

17. **Binder, M., Tourmente, S., Roth, J., Renaud, M., and Gehring, W. J.,** *In situ* hybridization at the electron microscope level: localization of transcripts on ultrathin sections of lowicryl K4M-embedded tissue using biotinylated probes and protein A-gold complexes, *J. Cell. Biol.,* 102, 1646, 1986.

18. **Morel, G., Dubois, P., and Gossard, F.,** Détection ultrastructurale des ARN messagers codant pour l'hormone de croissance dans lantéhypophyse du rat par hybridation *in situ, C. R. Acad. Sci. Paris,* 302, 479, 1986.

19. **Steinert, G., Felsani, A., Kettmann, R., and Brachet, J.,** Presence of rRNA in the heavy bodies of sea urchin eggs, *Exp. Cell Res.,* 154, 203, 1984.

20. **Tong, Y., Zhao, H. F., Simard, J., Labrie, F., and Pelletier, G.,** Electron microscopic autoradiographic localization of prolactin mRNA in rat pituitary, *J. Histochem. Cytochem.,* 37, 567, 1989.

21. **Webster, H. de F., Lamperth, L., Favilla, J. T., Lemke, G., Tesin, D., and Manuelidis, L.,** Use of a biotinylated probe and *in situ* hybridization for light and electron microscopic localization of Po mRNA in myelin-forming schwann cells, *Biochemistry,* 86, 441, 1987.

22. **Angerer, L. M., Stoler, M. H., and Angerer, R. C.,** *In situ* hybridization with RNA probes: an annotated recipe, in *In Situ Hybridization: Applications to Neurobiology,* Valentino, K. L., Eberwine, J. H., and Barchas, J. D., Eds., Oxford University Press, New York, 1987, chap. 3.

23. **Singer, R. H., Lawrence, J. B., and Rashtchian, R. N.,** Toward a rapid *in situ* hybridization methodology using isotopic and nonisotopic probes, in *In Situ Hybridization: Applications to Neurobiology,* Valentino, K. L., Eberwine, J. H., and Barchas, J. D., Eds., Oxford University Press, New York, 1987, chap. 4.

24. **Guitteny, A. F., Böhlen, P., and Bloch, B.,** Analysis of vasopressin gene expression by *in situ* hybridization and immunohistochemistry in semi-thin sections, *J. Histochem. Cytochem.,* 36, 1373, 1988.

25. **Angerer, L. M. and Angerer, R. C.,** Detection of poly A + RNA in sea urchin eggs and embryos by quantitative *in situ* hybridization, *Nucleic Acids Res.,* 9, 2819, 1981.

26. **Lawrence, J. B. and Singer, R. H.,** Quantitative analysis of *in situ* hybridization methods for the detection of actin gene expression, *Nucleic Acids Res.,* 15, 1777, 1985.

27. **Roth, J., Bendayan, M., Carlemalm, E., Villiger, W., and Garavito, M.,** Enhancement of structural preservation and immunocytochemical staining in low temperature embedded pancreatic tissue, *J. Histochem. Cytochem.,* 29, 663, 1981.

28. **Chesselet, M. F., Weiss, L., Wuenschell, C., Tobin, A. J., and Affolter, H. U.,** Comparative distribution of mRNAs for tachykinins in the basal ganglia: an *in situ* hybridization study in the rodent brain, *J. Comp. Neurol.,* 262, 125, 1987.

29. **Tecott, L. H., Eberwine, J. H., Barchas, J. D., and Valentino, K. L.,** Methodological considerations in the utilization of *in situ* hybridization, in *In Situ Hybridization: Application to Neurobiology,* Valentino, K. L., Eberwine, J. H., and Barchas, J. D., Eds., Oxford University Press, New York, 1987, chap. 1.

30. **Baldino, F., Chesselet, M. F., and Lewis, M. E.,** High-resolution *in situ* hybridization histochemistry, in *Neuroendocrine Peptide Methodology,* Academic Press, New York, 1988.

31. **Langer, P. R., Waldrop, A. A., and Ward, D. C.,** Enzymatic synthesis of biotin-labeled polynucleotides: novel nucleic acid affinity probes, *Proc. Natl. Acad. Sci. U.S.A.,* 78, 6633, 1981.

32. **Landegent, J. E., Jansen in de Wal, N., Baan, R. A., Hoeijmakers, J. H. J., and Ploeg, M. van der,** 2-Acetylaminofluorene-modified probes for the indirect hybridocytochemical detection of specific nucleic acid sequences, *Exp. Cell Res.,* 153, 61, 1984.

33. **Baschong, W., Lucocq, J. M., and Roth, J.,** "Thiocyanate gold": small (2-3 nm) colloidal gold for affinity cytochemical labeling in electron microscopy, *Histochemistry,* 83, 409, 1985.

Chapter 8

QUANTITATIVE ANALYSIS OF *IN SITU* HYBRIDIZATION USING IMAGE ANALYSIS

Arnold J. Smolen and Patricia Beaston-Wimmer

TABLE OF CONTENTS

I. INTRODUCTION

While various methods exist to quantify messenger RNA (mRNA) levels in tissue homogenates and extracts, they do not permit individual cellular measurements in a heterogeneous population of cells. On the other hand, the technique of *in situ* hybridization is the best method available for the localization of specific messages within individual cells. The use of video-based computer image analysis offers the potential for the development of such a technique, in which measurements of mRNA levels can be made within single cells in histological sections.

Several recent reviews have been published regarding the quantitative analysis of *in situ* hybridization.[1-5] The purpose of this chapter is to describe the conditions that are required to obtain quantitative data from *in situ* hybridization studies, at both the tissue and the cellular level, and to describe the image analytic procedures by which such quantitative measurements can be made rapidly and reliably.

A. Current Techniques for Measuring mRNA Levels

Several methods are currently available for the measurement of mRNA levels, including Northern blot and dot blot. For either of these procedures, the cells must be lysed and the RNA extracted.[6] For Northern blotting, the isolated RNA must be denatured, subjected to electrophoresis, and hybridized to a restriction fragment of the appropriate complementary DNA probe.[7-9] For RNA dot hybridization, different amounts of RNA are immobilized onto nitrocellulose, and again exposed to a nick translated fragment of the appropriate complementary DNA.[8] When radiolabeled cDNA is used, it is possible to measure the amount of mRNA by reading the density of the autoradiographic spots.

While these methods can yield very precise and accurate quantitative data regarding the amount of specific messages, they cannot produce any data regarding the cellular distribution of these messages within the block of tissue that is analyzed.

B. *In Situ* Hybridization for Localization of mRNA

In situ hybridization is a technique that permits the precise cellular (and subcellular) localization and identification of cells that express a particular nucleic acid sequence. The essence of the technique is the hybridization of a nucleic acid probe with a specific nucleic acid sequence found in a tissue section. The first experiments involving *in situ* hybridization were published nearly 20 years ago.[10-12] Since that time, *in situ* hybridization has been used in many fields, including genetics, developmental biology, and virology. Recently, *in situ* hybridization has become of great importance in neurobiology as a powerful method for localizing individual cells that contain a particular species of mRNA within the complex, heterogeneous substance of the nervous system.

II. TISSUE PREPARATION CONSIDERATIONS
FOR QUANTIFICATION

A. General Considerations

In order to obtain quantitative data from histological sections that are reacted for *in situ* hybridization histochemistry, it is especially important to follow a rigid protocol for all stages of tissue handling, from tissue preparation through radioautography. It is beyond the scope of the present work to discuss tissue preparation in detail. What is emphasized is that it is critical that all tissues that will be compared in subsequent analyses be treated identically in all phases of tissue preparation.

B. Detailed Description of Tissue Processing Methods

In the material described here, we have used a commercially available (duPont NEP-229) antisense synthetic oligonucleotide probe for tyrosine hydroxylase mRNA (TH-mRNA), and have used a modified version of the protocol described by Lewis et al.[13,14] The tissue that is studied is the superior cervical sympathetic ganglion (SCG) of the rat, a source rich in cells that synthesize TH. The oligonucleotide is a >95% full length sequence anti-sense synthetic oligonucleotide probe for TH-mRNA, and the 3'-end is "tailed" with [35]S.

The SCG is removed, rapidly frozen on powdered dry ice, and stored at −80°C until needed (up to 1 month). Cryostat sections (10 μm) are cut at −20°C and thaw mounted on subbed slides. The slides are kept at the bottom of the cryostat, and then are stored at −80°C until used. At the time of use, the slides are warmed to room temperature for 30 min and fixed by immersion in 3% buffered paraformaldehyde for 5 min at room temperature and rinsed briefly in phosphate buffered saline (PBS). The sections are incubated for 10 min in 2X SSC at room temperature (SSC, 1X) contains 0.15 M sodium chloride and 0.015 M sodium citrate at pH 7.0). Excess SSC is removed and the slides are placed on wet filter paper in a culture dish.

Sections are then incubated at 37°C in a humid environment for 1.5 h in prehybridization buffer, prepared as follows:

5 ml	deionized formamide
2 ml	20X SSC
0.2 ml	50X Denhardt's solution (Sigma)
0.5 ml	salmon sperm DNA (10 mg/ml)
0.25 ml	yeast tRNA (10 mg/ml)
2 ml	dextran sulfate (probe grade; Oncor Inc.)
15 mg	dithiothreitol (10 mM final concentration)

Hybridization is then carried out in the same buffer with 3.0×10^6 cpm of labeled probe per milliliter of hybridization medium (25 μl of medium is sufficient to produce a drop that completely covers the SCG section). After overnight (16 h) hybridization, the sections are washed in 2X SSC at room temperature for 2 h, 1X SSC at room temperature for 2 h, 0.5X SSC at room temperature for 1 h, 0.5X SSC at 37°C for 30 min (twice), and 0.5X SSC at room temperature for 30 min (all SSC washes contain 14 mM 2-mercaptoethanol and 1% sodium thiosulfate).

The slides are then dipped quickly in 300 mM ammonium acetate, and dehydrated in 70% ethanol with 300 mM ammonium acetate (1 min), 80% ethanol with 300 mM ammonium acetate (1 min), 95% ethanol (1 min), 100% ethanol (twice, 5 min each), xylene (5 min), xylene (30 min), and 100% ethanol (twice, 5 min each). The sections are then dried under a cool stream of air and stored at 4°C in a dessicator under vacuum overnight.

III. RADIOAUTOGRAPHIC CONSIDERATIONS
FOR QUANTIFICATION

A. General Considerations

The half-life and maximum particle energy of the nuclide are both important variables to consider in the selection of the appropriate choice of label for quantitative *in situ* hybridization histochemistry. The ideal radionuclide for use in radioautography should have a half-life that is neither too long (since high specific activities cannot be attained) nor too short (since the necessary exposure times may outlast the decay). Similarly, the emission energy should be neither too low (since particles will have difficulty escaping through the thickness of the section into the emulsion) nor too high (since particles will be less likely to collide with a silver halide crystal in the emulsion, and in addition, the exposed silver grains may lie at some distance from the sources of the radioactive decay).

B. Spatial Resolution

The spatial resolution provided by any radionuclide is inversely related to the maximum energy of the beta particles resulting from its decay.[15] ^{3}H-labeled probes, because of their low energy (maximum of 19 keV), provide excellent spatial resolution. [125]I does not release beta particles; rather it decays by electron capture, yielding particles of a variety of energies, most of which are less than 4 keV.[15] Therefore, the spatial resolution provided by ^{125}I-labeled probes is somewhat better than that of ^{3}H probes. Two problems result from the use of these low energy probes. One is that the particles that are emitted may not be able to traverse the thickness of the tissue section, resulting in a bias toward labeling of only the most superficial layers of the section. A second problem is that quenching, or differential absorption of the radioactive decay particles by various organelles, is more obvious than with radionuclides that emit low energy particles.

[14]C- and ^{35}S-labeled probes also provide acceptable cellular resolution at the

light microscopic level. The maximum beta particle energy resulting from decay of ^{14}C is 156 keV and of ^{35}S is 167 keV. ^{32}P probes do not provide acceptable cellular resolution, since the beta particle has very high energy (1709 keV). ^{32}P labeled probes are more useful for *in situ* hybridization experiments in which regional, but not cellular, resolution is required.

C. Exposure Times

The exposure time that is required for any labeled probe is directly related to the specific activity of the probe. Since there is an inverse relationship between the maximum specific activity of a radionuclide and its half-life,[15] radionuclides that have shorter half-lives can be used to produce probes of higher specific activities than can radionuclides with longer half-lives. ^{14}C has an extremely long half-life (5730 years), and ^{3}H also has a relatively long half-life (12.6 years). Therefore, ^{14}C- and ^{3}H-labeled probes generally have low specific activities and require long exposure times (weeks to months). Probes labeled with ^{32}P, which has a very short half-life (14.3 d), can attain very high specific activities, and thus result in very short exposure times (days). However, the short half-life of ^{32}P-labeled probes has the disadvantage that significant radioactive decay of the isotope occurs during the exposure time. Thus, the exposure time must be managed critically. ^{35}S has a half-life of 87.9 d and ^{125}I has a half-life of 60.0 d, so both radionuclides can produce labeled probes with high specific activities. Exposure times for ^{35}S- or ^{125}I-labeled probes are on the order of days to weeks. In addition, because of the longer half-life of ^{125}I or ^{35}S, compared to ^{32}P, the precise timing of the exposure is not as critical. Therefore, for *in situ* hybridization experiments in which cellular resolution is required, the most satisfactory probe is one labeled with ^{125}I or ^{35}S.

D. Other Considerations

The thickness of the emulsion is an important consideration in the quantitation of *in situ* hybridization experiments, particularly those using high energy probes, such as ^{32}P or ^{35}S. Since the path length of the beta particle resulting from the decay of these radionuclides is much longer than the thickness of the emulsion, variations in the emulsion thickness will result in variations in the number of exposed silver grains. Several techniques exist to minimize the variability in emulsion thickness.[15] A tightly packed monolayer of emulsion is generally regarded as ideal for quantitative radioautographic studies. Such a monolayer can be produced by critically standardizing the dilution of the emulsion and the rate of withdrawal of the slide from the emulsion.

The exposure time should be regulated such that saturation of the emulsion does not occur. That is, test slides should be developed at regular intervals to determine that no overlapping of exposed grains occurs over the most heavily labeled cells. Once such overlapping occurs, the area occupied by grains and the optical density cease to relate linearly to the number of grains, and the tissue is not useful for quantitative analysis.

E. Detailed Description of Radioautographic Method

Following the hybridization reaction (described above) and after the sections have been stored overnight in a dessicator at 4°C under vacuum to dry, the slides are brought to room temperature, and are coated with liquid emulsion. A relative humidity level of at least 80% is obtained in the darkroom by running hot tap water for at least 1 h. A solution of 600 mM ammonium acetate in distilled water is placed in a Coplin jar in a warm water bath (42°C). An equal volume of Kodak NTB-3 emulsion is placed in the Coplin jar (final ammonium acetate concentration is 300 mM), and the emulsion is allowed to melt for 20 to 30 min. The liquid emulsion is then stirred gently, and test slides are dipped at a slow, steady rate. The excess emulsion is drained from the slides, and the slides are kept in a vertical position to dry for 20 min.

The test slides are then developed as follows: 3.5 min in D-19, rinse in water for 1 min, and fix for 5 min in Kodak Rapid Fix (all at 15°C). After a 5-min wash in running water, the slides are examined under the microscope.

After it has been determined that the procedure is satisfactory, the experimental slides are dipped in the manner described above. The slides are allowed to dry for 3 h in the humid environment, and are then placed in boxes containing Drierite. The boxes are wrapped in clear plastic wrap and then in aluminum foil, and are stored at 4°C until development. Test slides are developed at 3-d intervals until the desirable exposure time is determined (high signal to background ratio, but individual silver grains are resolvable over the most heavily labeled cells). At this point, all of the experimental slides are developed and counterstained lightly with cresyl violet.

IV. IMAGE ANALYTIC TECHNIQUES

A. General Considerations

To obtain quantitative information regarding the amount of mRNA that is present in any particular cell, we developed a PC-based image analysis system, MORPHON. The MORPHON system consists of an MTI Dage Model 65 monochrome video camera, a PCVision Plus Frame Grabber (Imaging Technology, Inc.), a Sony analog RGB monitor, and a Numonics Model 2207 digitizing tablet connected to an 80286-based IBM-AT compatible computer (PC Designs). The MORPHON image analysis software is a comprehensive set of routines, written in Turbo Pascal (Borland), that perform a variety of image processing and image analytic manipulations.

It is critical in the acquisition of quantitative data from tissue sections that optimal and standardized observation conditions are routinely met. First, critical (Koehler) illumination is obtained through the microscope optics. Next, the illumination of the image is standardized by setting a blank region of the slide to a fixed gray level, e.g., 240 on a scale of 0 to 255, where 0 is black and 255 is white. For this step, it is imperative that the light source is stable and that all automatic features of the camera (gain, target, black level) are disabled. When these conditions are rigorously satisfied, it is possible to reproduce measurements from one session to the next with a high degree of reliability.

B. Tissue Level of Resolution

Once all of the above requirements for tissue preparation, histochemical reaction, and radioautography have been met, it is possible to obtain very reliable estimates of the amount of radioautographically labeled mRNA in discrete brain regions. For the tissue level of resolution, film radioautographs have several advantages over emulsion dipped radioautographs. The most important advantage is that, once developed, the film has no underlying tissue to confound the analysis. This advantage is also a disadvantage, since the lack of tissue can make recognition of regional boundaries very difficult.

The computer based image analysis system, however, can overcome this limitation as follows. First, the stained section from which the film was made is placed on the viewing box, and its image is displayed on the monitor. Using the digitizing tablet, the operator can trace outlines around whichever regions are desired for subsequent analysis, and can associate names with each of these traced regions. The outlines are displayed on the monitor superimposed over the tissue section.

When this step is finished, the stained section is removed, and the film is placed on the viewing box. To facilitate alignment of the film, the traced outlines remain on the monitor while the live image of the film is displayed. Once positioned, the computer is programmed to perform an average brightness measurement within each of the outlined regions.

After this is accomplished, the operator is prompted to superimpose a measuring box over a region of the film that is considered to be tissue background. The average brightness of this background region is then computed. Finally, the operator is asked to superimpose a measuring box over a region of the film that was not exposed to tissue, and its average brightness is obtained. The computer then calculates a relative optical density from the average brightness measurements for each of the outlined regions. The reference against which these relative measurements are made is the brightness of the blank region of the film (considered to be 100%).

First, the relative optical density of the outlined region of interest is calculated (B represents the brightness value, scaled 0 to 255):

$$OD_{region} = \frac{1}{\log(B_{region}/B_{blank})} \tag{1}$$

Next, the relative optical density of the tissue background is determined:

$$OD_{background} = \frac{1}{\log(B_{background}/B_{blank})} \tag{2}$$

Finally, the relative optical density of each region is divided by the relative optical density of the tissue background. In this way, the amount of label in each region is expressed as a ratio to background values. A ratio of 1.0 indicates that the amount of label in the region of interest is the same as that found in the background, while higher ratios indicate specific labeling of the region.

C. Cellular Level of Resolution

Measurements of mRNA labeling at the cellular level are more difficult, since only emulsion dipping provides the requisite resolution. Therefore, the final product consists of developed silver grains superimposed on the tissue section. Further complicating the analysis is the necessity for staining the tissue in order to identify the particular cells of interest.

Figures 1a and 1b illustrate the difficulty in resolving the silver grains from the underlying stained cytoplasm. Figure 1a is taken with the stained cell stained cell bodies in focus, and Figure 1b is of the same area, with the exposed silver grains in sharp focus. The black and white illustrations provide a good approximation to the images that are available to the computer, since nearly all image analysis systems currently in use employ a monochrome video camera for image acquisition.

The strategy described above for analysis at the tissue level of resolution fails at the cellular level since the computer will average the underlying stained tissue into the brightness measurements along with the silver grains. Even standard image processing techniques, such as segmentation by brightness, are unable to distinguish the silver grains from the underlying tissue, since the brightness values of these two components are quite similar.

More sophisticated image processing techniques are required in this instance to discriminate the silver grains. One alternative is to use dark field optics to produce brightly glowing silver grains over a relatively dark background. Rogers et al.[4] used this method to obtain automated grain counts of cells labeled with a probe for somatostatin mRNA. Their procedure was performed in two steps. First, cells of interest were mapped using bright field optics, and the coordinates of these cells were stored in the computer. Then the computer scanned the section and obtained grain counts using dark field optics. This approach has several limitations. The most significant is that it requires a computer-driven X-Y microscope stage.

Our approach is to use bright field optics with image processing to enhance the radioautographic grains. Preliminary experiments indicate that the greater the optical magnification employed, the more satisfactory the resulting analysis. One reason for this is that higher optical magnifications result in better spatial resolution, and another is that higher pixel to grain ratios are produced. Therefore, a 100X oil immersion objective was used for these studies.

The first step is to scan the area of interest, and identify cells to be measured. One at a time, the perimeters of the cells to be analyzed are traced (Figure 1c). The computer then performs image enhancement of the region within the tracing to sharpen the grains. This is accomplished by using a spatial convolution consisting of a 3 X 3 kernel:

$$\begin{matrix} -1 & -1 & -1 \\ -1 & 12 & -1 \\ -1 & -1 & -1 \end{matrix}$$

For each pixel in the area being analyzed, its brightness is multiplied by a factor of 12. Then each of the eight pixels that neighbor the center pixel is examined, and their values are each multiplied by a factor of −1. All nine values are summed, and center pixel takes on the value of the result.

This type of spatial filter is generally regarded to be a sharpening, or high pass filter, since it accentuates the high frequency components in the image. Because the silver grains are small, dark objects, they are sharpened using this filter, and stand out from the underlying stained cytoplasm. The cytoplasm is not enhanced because the transition from lighter areas to darker areas is gradual. The results of this image enhancement are illustrated in Figure 1d.

After the image has been filtered, the operator then sets an upper limit for density segmentation, so that the grains, which are now the darkest regions of the image, are selectively highlighted. The computer then measures the area of the grains and the total area of the outlined cell.

V. METHODS FOR VALIDATION

To validate the results of the quantitative data obtained by image analysis, we compared the resulting grain area calculated by the computer with manually performed grain counts on 50 labeled cells selected at random (Figure 2). The number of grains from the manual counts ranged from 10 to 75 per cell, and the calculated grain areas ranged from 6 to 51 $\mu m.^2$ Thus, a relatively wide range of values was examined. Figure 2 demonstrates that there is a close linear relationship between the computer generated grain areas and the manual grain counts. Linear regression analysis results in a correlation coefficient (r) of 0.935. These results indicate that the computer assisted quantification of grains described here, based upon automated area measurements, is a valid measure of grain numbers, and can be accomplished much more rapidly than is possible using manual techniques.

Several investigators have used standards to calibrate the amount of radioactivity in the tissue. In general, these standards are prepared by adding known amounts of isotope to brain homogenates. Samples of known weight of the homogenates are analyzed by liquid scintillation counting. The remainder of the homogenates are frozen and sectioned in the cryostat at the same thickness as the experimental tissue. These sections are then processed for radioautography along with the experimental tissue. From the series of homogenates, a standard

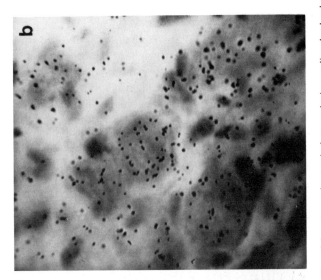

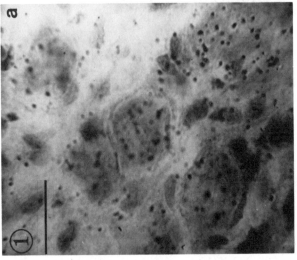

FIGURE 1. High power light micrographs of several cells from a 10-μm thick section of the rat superior cervical sympathetic ganglion that has been reacted for *in situ* hybridization using a ³⁵S-labeled synthetic oligonucleotide probe for tyrosine hydroxylase mRNA. (Bar represents 25 μm). (a) Several cresyl violet stained cells are visible in this illustration, and the developed silver grains are not in focus. It is difficult to distinguish the silver grains from the underlying tissue. (b) The plane of focus is now at the level of the exposed silver grains. In this illustration, the grains are more obvious, but they are still not sufficiently distinct from the underlying cells to be easily recognized by the computer. (c) The same section, after the operator has defined the perimeter of one cell. This boundary marks the area within which the semi-automated grain counts are made. (d) This illustrates the effects of image enhancement, using a sharpening or high pass spatial filter within the area to be measured. Note how the exposed silver grains now stand out from the underlying tissue, in comparison with surrounding regions where the enhancement has not occurred. At this point, the computer will measure the cross-sectional area occupied by silver grains and the total area of the cell, as defined by the traced boundary line.

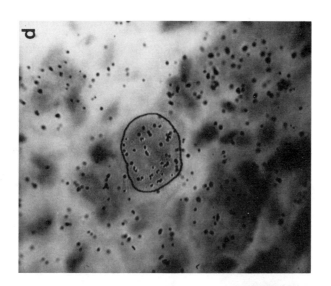

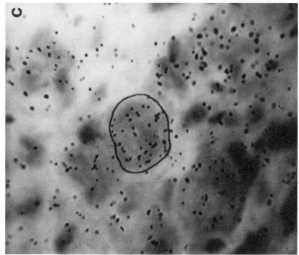

FIGURE 1 (continued).

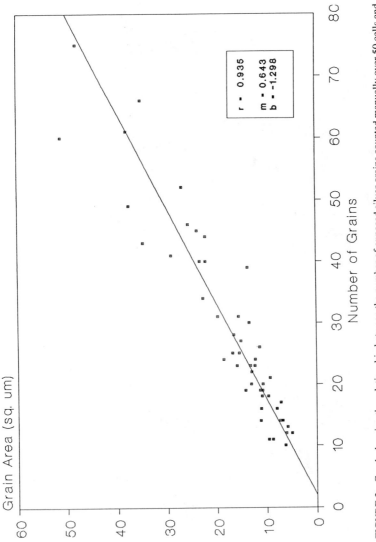

FIGURE 2. Graph showing the relationship between the number of exposed silver grains counted manually over 50 cells and the cross-sectional area occupied by grains, measured using the semi-automated image analysis system. There is a very good linear relationship (r = 0.935) of grain area to grain counts over a broad range of grain counts, from 10 to 75 grains per cell. Therefore, the measured grain area can be used as a valid predictor of the actual number of grains over each cell.

curve can be made that relates the grain density over a fixed area to a known amount of radioactivity. From the specific activity of the probe, it is then possible to estimate the actual number of molecules of probe, and thus the number of copies of the message in the cell. Interpretation of these data must be done with some caution, since many variables interact to produce the final grain density, including the efficiency of the hybridization, and the loss of mRNA during tissue preparation (see Young and Kuhar[5] for a detailed discussion of the use of standards in quantitative *in situ* hybridization studies). In addition, this method is limited to the use of film or emulsion-coated coverslips and cannot be used for sections dipped into emulsion.

An alternative approach[1] is to compare the amount of the specific mRNA under investigation with the total mRNA of the cell. To accomplish this, a series of adjacent sections are used, one set of which is hybridized with a specific probe, and the other is hybridized with a poly-U probe. Grain counts are then made in the adjacent sections of the same cells, and a ratio of specific mRNA to total poly-A mRNA is produced. While the information that is obtained is of significant value, the major drawback of the technique is that two separate hybridization reactions must be performed on the same tissue. In addition, this technique still does not permit an absolute quantitation of specific mRNA levels, since it is possible that certain experimental treatments may alter total poly-A mRNA levels as well as specific mRNA levels.

The methods described here allow reliable relative measurements to be made between tissue sections from the same hybridization reaction.[16] Comparisons between hybridization runs can perhaps be made most simply by the inclusion of a control series of sections in each hybridization reaction. In this way, the data from all of the experimental groups can be normalized against this internal control.

In summary, it is possible to use the technique of *in situ* hybridization to make rapid and reliable quantitative comparisons of specific mRNA levels at both the tissue and the cellular level, using carefully controlled experimental techniques along with rigorously implemented image analytic procedures.

REFERENCES

1. **Griffin, W. S. T.,** Methods for hybridization and quantitation of mRNA in individual brain cells, in *In Situ Hybridization: Applications to Neurobiology,* Valentino, K. L., Eberwine, J. H., and Barchas, J. D., Eds., Oxford University Press, New York, 1987, 97.
2. **Lewis, M. E., Krause, R. G., III, and Roberts-Lewis, J. M.,** Recent developments in the use of synthetic oligonucleotides for *in situ* hybridization histochemistry, *Synapse, 2,* 308, 1988.
3. **McCabe, J. T., Morrell, J. I., and Pfaff, D. W.,** *In situ* hybridization as a quantitative autoradiographic method: vasopressin and oxytocin gene transcription in the Brattleboro rat, in *In Situ Hybridization in Brain,* Uhl, G. R., Ed., Plenum Press, New York, 1986, 73.

4. **Rogers, W. T., Schwaber, J. S., and Lewis, M. E.,** Quantitation of cellular resolution *in situ* hybridization histochemistry in brain by image analysis, *Neurosci. Lett.,* 82, 315, 1987.

5. **Young, W. S., III and Kuhar, M. J.,** Quantitative *in situ* hybridization and determination of mRNA content, in *In Situ Hybridization in Brain,* Uhl, G. R., Ed., Plenum Press, New York, 1986, 243.

6. **Chirgwin, J. M., Przbyla, A. E., MacDonald, R. J., and Rutter, W. J.,** Isolation of biologically active ribonucleic acid from sources enriched in ribonuclease, *Biochemistry,* 18, 5294, 1979.

7. **Goldberg, D. A.,** Isolation and partial characterization of Drosophila alcohol dehydrogenase gene, *Proc. Natl. Acad. Sci. U.S.A.,* 77, 5794, 1980.

8. **Lewis, E. J., Tank, A. W., Weiner, N., and Chikaraishi, D.,** Regulation of tyrosine hydroxylase mRNA by glucocorticoid and cyclic AMP in a rat pheochromocytoma cell line. Isolation of a cDNA clone for tyrosine hydroxylase mRNA, *J. Biol. Chem.,* 258, 14632, 1983.

9. **Thomas, P. S.,** Hybridization of denatured RNA and small DNA fragments transferred to nitrocellulose, *Proc. Natl. Acad. Sci. U.S.A.,* 77, 5201, 1980.

10. **Buongiorno-Nardelli, S. and Amaldi, F.,** Autoradiographic detection of molecular hybrids between rRNA and DNA in tissue sections, *Nature,* 225, 946, 1970.

11. **Gall, J. G. and Pardue, M.,** Formation and detection of RNA-DNA hybrid molecules in cytological preparations, *Proc. Natl. Acad. Sci. U.S.A.,* 63, 378, 1969.

12. **John, H. A., Birnstiel, M. L., and Jones, K. W.,** RNA-DNA hybrids at the cytological level, *Nature,* 223, 582, 1969.

13. **Lewis, M. E., Sherman, T. G., and Watson, S. J.,** *In situ* hybridization histochemistry with synthetic oligonucleotides: strategies and methods, *Peptides,* 6(Suppl. 2), 75, 1985.

14. **Lewis, M. E., Arentzen, R., and Baldino, F., Jr.,** Rapid high-resolution *in situ* hybridization histochemistry with radio-iodinated synthetic oligonucleotides, *J. Neurosci. Res.,* 16, 117, 1986.

15. **Rogers, A. W.,** *Techniques of Autoradiography,* Elsevier, New York, 1979.

16. **Weiss, L. T. and Chesselet, M.-F.,** Regional distribution and regulation of preprosomatostatin messenger RNA in the striatum, as revealed by *in situ* hybridization histochemistry, *Mol. Brain Res.,* 5, 121, 1989.

APPENDIX

IN SITU HYBRIDIZATION USING RADIOLABELED RNA PROBES[1]

The radiolabeled probe is precipitated by ethanol, the pellet dissolved in a small amount of 0.1% DEPC-treated dH_2O, the amount of radioactivity checked on an aliquot, and the probe diluted to the desired final concentration of 0.1% DEPC-treated dH_2O.

Prehybridization Washes

All prehybridization steps are performed in plastic Coplin jars treated overnight with 0.1% DEPC and autoclaved. All solutions are made with 0.1% DEPC-treated dH_2O and autoclaved before use.

NOTE: Turn on oven to 50°C.

All Steps Under Mild Agitation

1. Frozen sections (10 μm) kept at -70°C and dry mounted on 2× subbed slides (Porcine Gelatin) are quickly warmed to room temperature under a stream of cool air and fixed in 3% paraformaldehyde in 0.1 *M* PBS, containing 0.02% DEPC for 5 min.
2. 0.1 *M* PBS — rinse (2× for 1 min).
3. 2X SSC* for 1 min twice.
4. 0.1 *M* triethanolamine pH 8.0 with 125 μl acetic anhydride per 50 ml (acetic anhydride should be added at the last minute) — 10 min.
5. 2× SSC for 1 min.
6. 0.1 *M* PBS — 1 min.
7. Tris/glycine 0.1 *M* pH 7.0 — 30 min.
8. 2× SSC for 1 min — twice.
9. 70% ETOH (DEPC dH_2O) — 1 min.
10. 80% ETOH (DEPC dH_2O) — 1 min.
11. 95% ETOH (DEPC dH_2O) — 1 min.
12. Air dry on rack - LABEL SLIDES.

Hybridization

Two pairs of gloves should be worn from this point on.

*   Hybridization mix (20 μl or more) containing the probe is put on each section, then covered with 2 layers parafilm (hybridization mix kept at −70°C in aliquots).
*   Hybridization in humid boxes for 3.5 to 4 h in a forced air oven set at 50°C.

*2X SSC is 0.15 *M* NaCl and 0.015 *M* sodium citrate.

Note:

1. 1 h before hybridization is complete, turn on water baths at 52°C and 37°C.
2. Defrost deionized formamide (stored at –20°C). Prepare selection 50% formamide in 2X SSC and warm up to 52°C.

   Post-hybridization washes:
   1. Let slides cool off to room temperature.
   2. Rinse off coverslips in 2X SSC at room temperture.
3. In 52°C water bath — 50% formamide in 5X SSC — 5 min — shake gently.
4. In 52°C water bath — 50% formamide in SX SSC — 20 min — DO NOT TURN OFF THE WATER BATH — see #8 below.
5. At room temperature — 2× SSC for 1 min — twice.
6. Slides are treated with RNAse (100 µg/ml); dilution as follows: 100 10 mg/ml stock solution in 10 ml of 2X SSC (1:100 dilution) for 30 min in a water bath set at 37°C.
7. Slides are placed in clean jars containing 2X SSC for 1 min — 2× or 3× at room temperature.
8. 2X SSC and 50% formamide at 52°C for 5 min.
9. 2X SSC for 1 min — twice — room temperature.
10. 2X SSC and 0.5% Triton X-100 — overnight on gentle shaker.

Next Day:

1. 2X SSC and 0.05% Triton X-100 — 2 quick rinses.
2. Quick dip in dH$_2$O containing 300 m*M* ammonium acetate.
3. 70% ETOH in 300 m*M* ammonium acetate — 1 min.
4. 80% ETOH in 300 m*M* ammonium acetate — 1 min.
5. 95% ETOH — 1 min.
6. 100% ETOH — 5 min — twice.
7. Xylene — 5 min.
8. Xylene — 30 min.
9. 100% ETOH — 5 min— twice.

AIR DRY — AUTOGRAPHY

Hybridization mix for *in situ* hybridization experiments (final concentration)

   40% formamide
   10% dextran sulfate
   1× Denhardt's solution
   4X SSC and 10 m*M* DTT
   1 mg/ml tRNA (coli)
   1 mg/ml denatured salmon sperm DNA

- For 2 ml: 1 ml prepared in each of 2 Eppendorf tubes

1. Combine   400 ml deionized formamide + 20% dextran sulfate
   200 ml 20X SSC
   100 ml tRNA (stock 10 mg/ml)
   100 ml DNA (stock 10 mg/ml)
2. Boil for 5 min and put on crushed ice immediately.
   THIS STEP IS CRUCIAL.
3. Eppendorf tube on ice:
   10 ml 1.0 M DTT (freshly thawed aliquot of DTT 1 M) (DTT must be kept on ice at all times after quickly thawing at room temperature) and 100 ml Denhardt's solution + BSA (RNAse-free).
4. Vortex each tube — centrifuge briefly and aliquot in 180 ml.
5. Store in –70°C freezer.

SYNTHESIS OF RADIOLABELED RNA PROBES

1. Turn on incubator at 39°C.
2. Take out all ingredients from –20°C freezer, put them on ice — final reaction volume 20 μl/tube.
3. Label tubes according to DNA to be used (on top and side of tube).
4. The following are added in the order listed (mix well by tapping between additions):

	Sense DNA	Notes
5X transcription buffer	4 μl	Stored at –20°C & brought to RT
100 mM DTT	2 μl	Stored at –20°C & brought to RT
RNAsin	1 μl	Stored at –20°C & brought to RT
10 mM GTP	1 μl	Stored at –20°C & brought to RT
10 mM ATP	1 μl	Stored at –20°C & brought to RT
10 mM CTP	1 μl	Stored at –20°C & brought to RT
100 μM UTP cold	2 μl	Stored at –20°C in conc. 10 mM UTP - dilute from stock soln. before use & discard remaining
(1 μg/μl) DNA (linearized)	2 μl	0—4°C storage
^{35}S-UTP	5 μl	–70°C storage
SP6 RNA Polymerase (or other appropriate polymerase)	1 μl	–20°C storage (keep on ice at all times while using and remove from the 20° only for theshortest time necessary)
TOTAL VOLUME	20 μl	

Mix well - *Centrifuge briefly* - Incubate 1 h at 39°C.

Optional: Add 1 µl more of SP6 Polymerase after 1 h and let the reaction continue for one more hour.

Add:

RQ DNAse	1 µl	(–20°C storage) To destroy DNA template
RNAsin	1 µl	RNase - inhibitor

Incubate: 15 min at 37°C (should be accurate)

Add:

100X EDTA (100 m*M*)	10 µl	(mix well)
10X STE	10 µl	(mix well)

Add:

Adjust vol. to 100 µl with DEPC dH$_2$O	58 µl	58 µl
	100 µl	100 µl

EXTRACT WITH PHENOL: CHLOROFORM

NOTE: Phenol is stored in 500 µl aliquots in a –20°C freezer. Defrost at 60°C and equilibrate by adding 500 µl TE buffer pH 7.6 (stored 4°C). Mix well and centrifuge 5 min. Chlorform/isoamyl alcohol 24:1 is stored at room temperature.

		Sense	**Antisense**
Step 1	ADD:	50 µl Phenol	50 µl Phenol
		Vortex	
		Let sit 5 min	Let sit 5 min
		Centrifuge 4 min	Centrifuge 4 min

Remove and discard bottom organic phase with pipette set for vol. of phenol that was initially added (50 µl) (RNA probe is in the top aqueous phase).

Step 2	ADD:	50 µl Phenol	50 µl Phenol
		50 µl Chloroform/ isoamyl alcohol	50 µl Chloroform/ isoamyl alcohol
		Vortex	Vortex
		Let sit 5 min	Let sit 5 min

| Centrifuge 4 min | Centrifuge 4 min |

Remove and discard bottom organic phase with pipette set for approximately 100 μl.

Step 3   ADD:

50 μl Chloroform/	50 μl Chloroform/
isoamyl alcohol	isoamyl alcohol
Vortex	Vortex
Let sit 5 min	Let sit 5 min
Centrifuge 4 min	Centrifuge 4 min

Remove and discard bottom phase approximately 50 μl.

ETOH PRECIPITATION

Purpose: To precipitate the RNA probe from the aqueous phase remaining after the phenol chloroform extraction. This allows you to concentrate your RNA and resuspend it in the appropriate solution to go on to the next step. It also eliminates the nonincorporated nucleotides.

Reagents:

1. 100 % ETOH kept in –20°C freezer and reserved for this use.
2. 4.5 M NaOAc pH 6.0.
3. tRNA (10 mg/ml) (kept in aliquots at –20°C).

Procedure:

1. Add 1:10 4.5 M NaOAc pH 6.0 (i.e., if your vol = 100 μl 1/10×/100 μl = 10 μl NaOAc.
2. Add 3 × vol ice cold ethanol.
3. 1 μl of 10 mg/ml of *E. coli* tRNA.
4. Mix well by inversion and tapping the tube.
5. Centrifuge briefly 1—2 s.
6. Let chill on dry ice mixed with ethanol for 30—60 min or overnight in –70°C freezer to allow precipitate to form.
7. Let come back briefly to room temperature until liquid.
8. Centrifuge for 30 min at 12,000 × g (maximal speed).
9. Pour supernatant into labeled tubes and save these in case all the RNA probe is not precipitated and it is necessary to recentrifuge (should be kept on ice).
10. After pouring the supernatant off tubes must remain inverted until all excess moisture is evaporated. Remove carefully remaining liquid with ***sterile*** Q-Tips.
11. Resuspend pellet in 20 μl DEPC dH$_2$O with 20 mM DTT (freshly made from defrosted aliquot of 1.0 M DTT).

IN SITU HYBRIDIZATION HISTOCHEMISTRY WITH RADIOACTIVE OLIGONUCLEOTIDE PROBES[2]

Labeling Reaction

Tailing of 35 pmol of an oligonucleotide probe at the 3′ end is accomplished using a commercially available DNA tailing kit (Boehringer Mannheim Cat. No. 1028 707).

1. To a sterile 1.5 ml Eppendorf tube, add:

> 2 μl   tailing buffer (vial 1)*
> 3 μl $CoCl_2$ solution (vial 2)*
> 1—2 μl   oligonucleotide probe
> 9.0 μl   [^{35}S]dATP (DuPont NEN, #NEG-034H)
> 1.0 μl   55 U terminal transferase (vial 4)
> Adjust vol to 25 μl with HPLC grade sterile water

* Tailing Buffer: 200 mM K cacodylate; 50 mM Tris HCl pH 6.6; 2 mM DTE; 500 μg/ml BSA; 1.5 mM $CoCl_2$.

Mix gently by flicking tube several times with finger; do not vortex. Briefly microfuge tube and incubate at 37°C for 5 min.

2. At the end of the reaction, unincorporated nucleotides can be separated using a NENSORB 20 (DuPont NEN Products #NLP-022) column as follows:

    a. Settle each column (one per reaction) by tapping side, until all column matrix is at the bottom.
    b. Remove cap and attach column to ring stand over disposable beaker.
    c. Fill with 3 ml absolute methanol (HPLC grade if possible), force through column with 10 ml syringe attached to column adaptor.
    d. Fill with 3 ml 0.1 M Tris HCl pH 8.0, force through with syringe.
    e. Add 500 μl Tris to reaction mixture, mix by pipetting, and add to column.
    f. Apply gentle pressure to the syringe (1 drop/2 s) until solution is through.
    g. Wash column with 1.5 ml Tris, force through with syringe.
    h. Elute oligonucleotide by adding 0.5 ml of 20% ethanol (in RNase-free H_2O); apply gentle pressure to syringe and slowly collect 12 drops to sterile Eppendorf tube.
    i. Count aliquot by liquid scintillation counting; should get about 500,000 cpm/μl.
    j. Add 1:10 vol of freshly prepared 100 mM dithiothreitol (final conc = 10 mM) for ^{35}S probes only.

Tissue Preparation & Pre-Hybridization Washes

Fresh frozen sections (12 μm) are sectioned on a crystat and thaw-mounted on 2X subbed slides (porcine gelatine) and stored –80°C. Immediately before the experiment the sections are quickly brought to room temperature under a stream of cool air and fixed in:

1. 3% Paraformaldehyde in 0.1 M PBS for 5 min.
2. Rinse 3× for 5 min in 0.1 M PBS.
3. Rinse 1× for 10 min in 2X SSC.
4. Optional: Slides are placed in a humid chamber and approximately 300 μl of hybridization buffer (see below) is pipetted onto each slide. Slides are permitted to incubate at RT for 1 h.

Hybridization buffer is prepared fresh by adding:

> 5 ml deionized formamide
> 2 ml 20X SSC
> 0.2 ml 50X Denhardt's
> 0.5 ml 10 mg/ml salmon sperm DNA*
> 0.25 ml 10 mg/ml yeast tRNA
> 2 ml 50% dextran sulfate

* DNA is denatured immediately before use by heating to 95°C for 10 min.

Hybridization

1. The radioactive probe is diluted with prehybridization buffer, allowing 500 μl of buffer per slide with 1.5 million cpm of activity (6 brain sections/slide). A higher concentration of probe may be helpful in some cases. If desired, parafilm coverslips can be applied to reduce volume.
2. Slides are then incubated at 37°C overnight (this temperature can be increased to 42°C for oligonucleotides greater than 36-mer).

Post-Hybridization

1. The slides are washed as follows:

    a. 2X SSC for 1 h at RT.
    b. 1X SSC for 1 h at RT.
    c. 0.5X SSC for $^1\!/_2$ h at 37°C.
    d. 0.5X SSC for $^1\!/_2$ h at RT.

NOTE: For ^{35}S-labeled probes add 2-mercaptoethanol, 14 mM final, and sodium thiosulfate, 1% final, to all washes.

2. The slides are rapidly dried by blowing a gentle stream of dried compressed air over them (through a Drierite column).
3. The dried slides are placed into an X-ray cassette with Kodak XAR-5 film. After developing the film (3 d exposure is a good starting point), the slides can be dipped in liquid emulsion.

GUIDELINES FOR AUTORADIOGRAPHY

1. *Preparation of the Darkroom*

— Check that there is enough water in the water bath to reach the level of your emulsion.
–– Turn on the water bath –42°C.
— Turn on the humidifier or let the water run in the sink to increase humidity up to at least 80—85%.
— Leave water running for duration of autoradiography.
— Check that all the glassware is *very clean,* in particular for traces of old emulsion.
— Make ready a sheet of aluminum foil to cover drying sections. Do not tear aluminum foil, plastic wrap, or tape in dark, as this will make sparks.

2. *Preparation of the Emulsion*

— Check the temperature of the water bath with a thermometer: must be 42°C.
— Check the humidity level: must be close to 80% at least.
— Add an equal volume (1:1 emulsion to AmAc-H_2O) of water containing ammonium acetate (final concentration 300 mM) to the Coplin jar in which you will melt the emulsion.
— Turn all the lights off (including red lights).
— Check that there is no light leak, in particular at the door: move slightly if necessary.
— Unfold the aluminum foil surrounding the box containing the emulsion. Open the box - take out the bottle. Scoop out the necessary amount (1:1). Put the bottle back and close the box so that it is light tight - DO NOT WRAP IN FOIL INSIDE THE DARKROOM: sparks generated by aluminum foil can contaminate the emulsion in the jar.
— *Cover* the jar with glass lid or parafilm.
— Cover the water bath with foil or lid.
NB: — it is better to wear gloves when touching the emulsion.
— DO NOT FORGET TO PUT A *CLEAR* SIGN ON THE DOOR (with *DATE!*): DO NOT ENTER.

- Put back unused emulsion in the refrigerator: wrap it in foil and label it and indicate approximate amount used and date on aluminum foil.
- Let emulsion melt approximately 20—30 min before dipping the slides.

3. *Test Slides*

- Have your developer, water, and fix in well-identified jars at +15°C. Keep the jar *covered* when not in use.
- Check that the door is light tight. Adjust it.
- Stir *very gently* the emulsion in the water bath with a *CLEAN* glass slide.
- Dip *slowly clean* glass slides (2) in emulsion.
- Let the test slides dry for 20 min in the vertical position. For safety, cover the rack loosely with aluminum foil.
- Cover water bath when done dipping test slides.
- Do not forget to put sign back on the door.

4. *Development of Test Slides*

- Check that the door is light tight.
- Develop test slides as usual (3.5 min in developer, wash in water for 5 min in fix). Rinse these a few minutes (5—10) under tap water.
- Examine under the microscope at 40× and 100×. Make sure the emulsion is in focus.
- *Remove* developer and fix from the darkroom.

5. *Dipping of the Slides*

- Order your slides in the order of development before going into darkroom.
- In the darkroom: check that the door is light tight.
- Dip each slide in the emulsion (kept in the water bath at all times). Count 10 s from the edge of the jar to the bottom. Stop. 10 s from the bottom to the edge of the jar. Let the drop back into the jar by putting the slide at an angle to the edge of the jar. Put the slide in the corresponding hole in the rack.

THIS ROUTINE MUST BE FOLLOWED RIGOROUSLY with as much regularity as possible. The goal is to obtain a uniform mono-layer of emulsion TAKE YOUR TIME. DO NOT RUSH. RELAX.

— *Loosely* cover the rack with foil (prepared ahead of time).
— *Do not* turn off the water bath: could generate sparks.
— Do not forget to put back the sign on the door.

6. **Preparation of the Black Boxes**

— Make sure the boxes are very clean and dust free. If you need to wash them: dry them very carefully (water tends to stay in the s holders). Use the air blower if necessary.
— Put a fair amount of dry Dierite wrapped in a kim wipe and hold it with a clean slide.
— Close the boxes once they have the dessicant in it.
— Prepare as many sheets of Saran Wrap and foil as you need for groups of boxes that will be developed together.

7. **Putting Slides in the Boxes**

— Allow the slides to dry for about 3 h in the *high humidity* room. Leave water or humidifier on to maintain humidity.
— Check that the door is light tight.
    Put no more than 3—4 slides per black box.
** CLOSE ALL THE BOXES before wrapping them. Saran Wrap and foil generate sparks!
— Turn the light on.
— Turn off the water or humidifier and the water bath.
— Clean up the mess.
— Clean up the jar that contained the emulsion *immediately* with water: do not allow the emulsion to cool down and dry in it. It be very difficult to clean up.

NONRADIOACTIVE *IN SITU* HYBRIDIZATION HISTOCHEMISTRY WITH OLIGONUCLEOTIDE PROBES USING CEPHALON PROTOCOL LABELING REACTION[3]

Labeling Reaction
Tailing of 35 pmol of an oligonucleotide probe at the 3′ end is accomplished using the BMB DNA Tailing Kit (Cat. No. 1028 707).

1.   To a sterile 1.5 ml Eppendorf tube, add:

Tailing buffer (vial 1)	4 µl
CoCl$_2$ solution (vial 2)	6 µl
Digoxigenin-11-dUTP*	2.5 µl
Oligonucleotide	1 µl

dATP (vial 5)	2 µl of 1/50 dilution**
Water	3.5 µl
TdT (vial 4)	1 µl

* Cat. No. 1093 088.
** 1/50 dilution is made in sterile water.

The reaction is incubated at 37°C for 5 min. After the 5 min incubation period, the tailed oligonucleotide is purified from the labeling reaction by ethanol precipitation.

Procedure: Adjust vol to 100 µl with sterile water
Add 1:10 (10 µl) 4.5 M Na acetate pH 6.0
3× vol ice cold ethanol (300 µl)
1 µl of 20 mg/ml glycogen (vial 9)
Mix by inversion and tapping the tube
Centrifuge briefly: 1—2 s
Let chill on dry ice mixed with ethanol for 30 min
 or overnight at –80°C
Centrifuge for 30 min at 12,000 × g (max. speed)
Pour off supernatant, keeping tube inverted
Allow pellet to dry and resuspend pellet in
 20 µl sterile water

Tissue Preparation & Pre-Hybridization Washes

Fresh frozen sections (12 µm) are sectioned on a cryostat and thaw-mounted on 2X subbed slides (porcine gelatin) and stored at –80°C. Immediately before the experiment the sections are quickly warmed to room temperature and fixed in:

1. 3% Paraformaldehyde in 0.1M PBS (plus 0.02% DEPC) for 5 min
2. 0.1 M PBS - Rinse 3× for 5 min.
3. 2X SSC - Rinse 1× for 10 min.
4. Slides are placed in a humid chamber and approximately 300 µl of hybridization buffer (see below) is pipetted onto each slide. Slides are permitted to incubate at RT for 1 h.

Hybridization

1. The digoxigenin-labeled probe is diluted with hybridization buffer to a volume of 1 ml. The required concentration of probe will depend upon mRNA abundance and must be determined experimentally.
 Hybridization buffer is prepared fresh by adding:

> 5 ml Deionized formamide
> 2 ml 20X SSC
> 0.2 ml Denhardt's
> 0.5 ml 10 mg/ml salmon sperm DNA*
> 0.25 ml 10 mg/ml yeast tRNA
> 2 ml dextran sulfate

* DNA is denatured immediately before use by heating to 95°C for 10 min.

2. Excess prehybridization buffer is removed from the slides by a quick dip in 2X SSC and the glass surrounding the tissue is carefully dried with a kim wipe. The probe is then applied allowing 30 µl per section after which a parafilm coverslip is applied. The slides are then incubated at 37°C overnight (this temperature can be increased to 42°C for oligonucleotides greater than 36-mer).

Post-Hybridization

1. The slides are washed (with gentle shaking) as follows:

> 2X SSC for 1 h at RT
> 1X SSC for 1 h at RT
> 0.5X SSC for 0.5 h at 37°C
> 0.5X SSC for 0.5 h at RT

Immunological Detection

1. Wash slides for 1 min in Buffer #1 at RT.
2. Incubate sections with 2% Normal Sheep Serum plus 0.3% Triton X-100 in Buffer #1 for 30 min at RT.
3. Dilute antibody conjugate (1:500) with Buffer #1 containing 1% Normal Sheep Serum and 0.3% Triton X-100. Pipette diluted anti-digoxigenin antibody conjugate onto sections and incubate at RT for 3—5 h in a humid chamber. Do not allow sections to dry out.
4. Wash slides for 10 min in Buffer #1 with shaking.
5. Washslides for 10 min in Buffer #2 shaking.
6. Incubate slides with approximately 500 µl/slide of color solution. Slides are placed in light-tight boxes on wetted filter paper backing. Slides can be checked periodically for color development (0.5—24 h).
7. Reaction can be stopped in Buffer #3.
8. Dehydrate sections as follows: 1 min in 70% ethanol, 1 min in 80% ethanol, 1 min in 95% ethanol, 1 min in 100% ethanol, 3 min in xylene. Slides can then be coverslipped in Permount (66% permount/33% xylene) mounting medium.

Solutions

Buffer #1 — 100 mM Tris-HCl; 150 mM NaCl; pH 7.5
Buffer #2 — 100 mM Tris-HCl; 100 mM NaCl; 50 mM MgCl$_2$; pH 9.5
Buffer #3 — 10 mM Tris-HCl; 1 mM EDTA; pH 8.0

Color Solution — 45 μl NBT solution
                         35 μl X-phosphate solution
                         2.4 mg levamisole
are added to 10 ml of Buffer #2 immediately before use.

IN SITU HYBRIDIZATION HISTOCHEMISTRY WITH NONRADIOACTIVE cRNA PROBES[3]

The synthesized RNA probe is ethanol precipitated and then diluted in 100 μl of 0.1% DEPC-treated water and stored on ice until hybridization.

Pre-Hybridization Washes

All pre-hybridization steps are performed in plastic Coplin jars treated overnight with 0.1% DEPC-treated water and autoclaved before use. All solutions are made with 0.1% DEPC-treated water and autoclaved before use. All steps are performed under mild agitation.

Fresh frozen sections (10 μm) are sectioned on a cryostat, thaw-mounted on 2X subbed slides (porcine gelatin), and stored at –80°C. Immediately before the experiment the sections are quickly warmed to room temperature under a stream of cool air and fixed in:

1. 3% Paraformaldehyde in 0.1 M PBS (plus 0.02% DEPC) for 5 min.
2. 0.1 M PBS - Rinse 2× for 1 min.
3. 2X SSC - Rinse 2× for 1 min.
4. 0.1 M Triethanolamine pH 8.0 with 125 μl acetic anhydride per 50 ml (acetic anhydride should be added at the last minute) - 10 min.
5. 2X SSC - Rinse 1× for 1 min.
6. 0.1 M PBS - Rinse 1× for 1 min.
7. 0.1 M Tris/glycine pH 7.0 - 30 min.
8. 2X SSC - rinse 2× for 1 min.
9. 70% ETOH (DEPC water) - 1 min.
10. 80% ETOH (DEPC water) - 1 min.
11. 95% ETOH (DEPC water) - 1 min.
12. Air dry on rack.

Hybridization: Label slides, apply 25 μl of diluted probe to each section, cover with 2 layers parafilm and hybridize in humid boxes for 3½ to 4 h. at 50°C.

Dilute probe by adding 900 μl of hybridization mix (stored in aliquots –80°C). Hybridization mix (final concentration):

> 40% Formamide
> 10% dextran sulfate
> 1X Denhardt's
> 4X SSC + 10 mm DTT
> 1 mg/ml tRNA (yeast)
> 1 mg/ml denatured salmon sperm DNA

Post-Hybridization Washes: All steps under mild agitation.

1. Let slides cool to room temperature.
2. Remove coverslips in 2X SSC. Place slides in 2X SSC RT.
3. In 52°C water bath — 50% formamide in 2X SSC - 5 min.
4. In 52°C water bath — 50% formamide in 2X SSC - 20 min.
5. At room temperature — 2X SSC — 2× for 1 min.
6. At 37°C water bath — RNase wash 100 μg/ml in 2X SSC — 30 min.
7. 2X SSC - 2× or 3× for 1 min.
8. At 52°C water bath — 50% formamide in 2X SSC - 5 min.
9. 2X SSC & 0.05% Triton X-100 containing 2% Normal Sheep Serum — overnight on gentle shaker.

Next Day:

1. Wash slides 2× for 5 min in Buffer #1 (100 mM Tris-HCl pH 7.5, 150 mM NaCl).
2. Incubate in anti-digoxigenin-AP conjugate (diluted 1:1000 in Buffer #1) containing 1% Normal Sheep Serum and 0.3% Triton X-100 for 5 h at room temperature.
3. Wash slides 1× for 10 min in Buffer #1.
4. Wash slides 1× for 10 min in Buffer #2 (100 mM Tris-HCl, pH 9.5, 100 mM NaCl, 50 mM MgCl$_2$).
5. Incubate slides with 500 μl/slide of Chromagen (prepared by adding 45 μl NBT solution, 35 μl X-phosphate solution, 2.4 mg levamisole to 10 ml Buffer #2) in a light-tight box on buffer-wetted filter backing. Slides must be kept in the dark during incubation (2—24 h), but can be checked periodically for color development.
6. Reaction can be stopped in 200 ml of Buffer #3 (10 mM Tris-HCl, pH 8, 1 mM EDTA).
7. Tissue can then be quickly dehydrated, cleared in xylene, and coverslipped.

PARTIAL ALKALINE HYDROLYSIS OF cRNA PROBE TO MAKE SHORT SEGMENTS[4]

Some investigators have suggested that partial alkaline hydrolysis of the probe into fragments of 100—150 bases greatly improves the signal, probably by improving probe penetration. This will have to be determined by experiment with the length of the cRNA probe being used.

After synthesis of your cRNA probe and resuspension in 100 µl of 20 mM DTT:

Take 50 µl:
1. Add 30 µl of 0.2 M Na_2CO_3 (Na carbonate)
   20 µl of 0.2 M $NaHCO_3$.
   Both DEPC-treated and autoclaved.
2. Mix well by tapping, centrifuge briefly, and incubate for calculated time in minutes (see below) at 60°C — *MUST BE PRECISE.*
3. Add immediately *on ice:*
   2.5 µl of Na acetate pH 6
   5 µl of 10% glacial acetic acid
4. Ethanol precipitate by adding:
   Add 7.5 µl of Na acetate pH 6
   300 µl of ice cold ethanol - 100%
   Place at –80°C O/N & centrifuge 30 min
5. Pour off supernanent & allow pellet to dry.
6. Resuspend pellet in 100 µl DEPC-treated H_2O.

NOTE: Calculated time can be determined by using the following formula: $L_o - L_f/KL_oL_f$, where L_o and L_f are the initial and final fragment lengths in kilobases (kb) and K is the rate constant for hydrolysis (0.11 kb/min).

REFERENCES

1. **Chesselet, M.-F., et al.,** *J. Comp. Neurol.,* 242, 125, 1987.
2. **Baldino, F., et al.,** *Methods Enzymol.,* 168, 761, 1989.
3. **Lewis, M. and Baldino, F.,** chapter 1.
4. **Cox et al.,** *Dev. Biol.,* 101, 485, 1984.

INDEX